TALKING STRAIGHT

TALKING STRAIGHT

LEE IACOCCA

WITH SONNY KLEINFIELD

BANTAM BOOKS
TORONTO • NEW YORK • LONDON • SYDNEY • AUCKLAND

TALKING STRAIGHT
A Bantam Book / June 1988

All rights reserved.
Copyright © 1988 by Lee Iacocca.
No part of this book may be reproduced or transmitted
in any form or by any means, electronic or mechanical,
including photocopying, recording, or by any information
storage and retrieval system, without permission in
writing from the publisher.
For information address: Bantam Books.

Library of Congress Catalog Card No. 88-47679

ISBN 0-553-05270-5

Published simultaneously in the United States and Canada

Bantam Books are published by Bantam Books, a division of Bantam Doubleday
Dell Publishing Group, Inc. Its trademark, consisting of the words "Bantam Books"
and the portrayal of a rooster, is Registered in U.S. Patent and Trademark Office
and in other countries. Marca Registrada. Bantam Books, 666 Fifth Avenue, New
York, New York 10103.

PRINTED IN THE UNITED STATES OF AMERICA

BOMC offers recordings and compact discs, cassettes
and records. For information and catalog write to
BOMR, Camp Hill, PA 17012.

To the 6.5 million people who bought
my first book . . .

and helped strike a blow against diabetes

CONTENTS

Acknowledgments	ix
A Brief Word	xi
PRELUDE: FIRED AGAIN	1
HEART AND HEARTH	19
I My Kids	21
II Growing Older	37
III Fame and Faith	53
PLAYING FOR BROKE	71
IV Good Business — More on Management	73
V Bad Business — What's Wrong with Wall Street	91
VI Business Sense — and Nonsense	107
VII I'll See You in Court	131
VIII The Press — We Can't Live Without It	143
IX Washout in Washington	163
SCANDAL, SCANDAL EVERYWHERE	181
X Free Trade or Free Ride	183

XI	Budget Busters	203
XII	The Food Crisis	219
XIII	The School Crisis	231
XIV	In Search of Quality	249

OUT FROM UNDER 265
    XV Twenty and Eight 267
    XVI If I Were President 283
    XVII Into the Twenty-first Century 301

Index 311

ACKNOWLEDGMENTS

Four years have gone by since I wrote my last book. You'd think the second time around would be a piece of cake, but it wasn't. Many people helped me with the first one. I needed even more help on the second.

I'd like to acknowledge some of those who pitched in. Again, I have to start with my absolutely indispensable secretaries. I used to have two; now I'm up to three and a half. That's because the mail has doubled in volume. Bonnie Gatewood, my number one secretary, runs the office and keeps me and my hectic schedule straight. She's the world's best, ably assisted by Valerie Kiss, Linda Gasparovich, and Sharon Gudenau.

Lots of other Chrysler people helped me get my facts straight. I am most grateful to Tom Denomme, Ben Bidwell, Bill O'Brien, Don Hilty, and Wally Maher. I can't forget Chris Pikulas and Lynn Feldhouse, who, along with me, read every letter that came in as a result of my first book. Mike Morrison, my speechwriter and confidant, kept me coherent and concise, as he usually does; and Tony Cervone, our resident Henny Youngman, supplied me with his weekly one-liners on the business world.

Wes and Isla Small helped me survive two weddings as the father of two brides in a year and a half. And other friends, such as Bill Winn, Hank Carlini, Bob Brown, Jay Dugan, Tom Clark, Carroll Shelby, Vic Damone, Bill Fugazy, Glenn White, and Dr. Ivan Mader, just helped me survive.

The Bantam crowd was even more professional and more fun the second time around. That includes Alberto Vitale, Linda Grey, Stuart Applebaum, Nina Hoffman, Steve Rubin, and Heather Florence.

I must especially thank Nessa Rapoport, my editor, not only for keeping my syntax straight but also for encouraging me to keep writing so that I could meet her tough deadlines.

Lastly, my deepest thanks to my collaborator, Sonny Kleinfield, who tolerated my ramblings and patiently put my words into a form of plain English that most people could understand.

Since writing does not come easily to me, and I had to spend hundreds of hours in seclusion, I want to thank my mom, Peggy, Kathi, and Lia for being so tolerant.

A BRIEF WORD

After writing one book, I knew enough about sequels to quit while I was ahead. *Rocky II* wasn't as good as *Rocky* I. *Jaws II* wasn't as good as *Jaws* I. Even *Meatballs II* wasn't as good as *Meatballs* I. And the movie's never as good as the book anyway, or is it the other way around?

I hate to admit it, but it's the same story in the auto business. Mustang I was a classic and loved by almost everyone. Mustang II laid a gigantic egg.

So when my publisher suggested that I write a second book, I said, "Wait a minute, I already wrote an autobiography. I regret I have only one life to give to my publisher. What on earth am I going to talk about?"

Rather than make a snap decision, I bounced the idea off some of my best friends. They were unanimous in their advice: "Write another book? Are you crazy?"

Even when I told my daughters that I might try my hand at a second book, they fought off yawns. Neither of them jumped up and down and said they couldn't wait to read it.

It was clear that my friends and family weren't dying to hear more words of wisdom from me. Nor did I have that much free time on my hands. True, once I finished raising money for the Statue of Liberty, I suddenly found I had a couple of free hours open between one and three A.M. But then I killed some time by buying American Motors, Lamborghini, and a piece of Maserati. I even bought an airplane company. I did another dozen television commercials and appeared in a forgettable episode of *Miami Vice*. I gave six commencement addresses, accepted five honorary degrees, wrote forty-two columns, and delivered a hundred and sixteen speeches. To fill up the weekends, I married off both my daughters. I occupied some more dead time by getting married myself, filing for divorce, reconciling, and ultimately getting divorced. I also bought a villa in Italy as a lark and got into the wine and olive oil business. And, unfortunately, I found time to gain twelve pounds.

But maybe the most important thing I did during the last four years was read 71,412 letters from you. That outpouring in response to my first book—in eighteen languages, from Sanskrit to Portuguese—truly amazed and humbled me. It was not only the most incredible thing that happened to me in the last four years, but the most incredible thing in my whole life.

I was so touched by all that mail that I decided to do my equivalent of 71,412 letters in one lengthy reply. I hope you're ready for it. It starts on the next page.

PRELUDE: FIRED AGAIN

A funny thing happened to me since I last wrote a book. I got fired. Again.

Fortunately, it wasn't from my real job this time. And Henry Ford didn't have a thing to do with it. It was the Washington politicians who tossed me out, as the chairman of the Statue of Liberty–Ellis Island Centennial Commission, after four years of hard work.

When I was given the news, I thought to myself: *This has got to stop. People are going to start thinking I'm a drifter, that I can't hold down a job!*

I was informed, would you believe, by telegram. It arrived on a Friday night in February 1986, just as I was about to go to bed. In classic Washington double talk, the wire praised me for my past efforts and then, in the best "we regret to inform you" style, told me there was the "potential of future conflict or the appearance thereof" between my roles as chairman of both the Statue of Liberty–Ellis Island Foundation, which had to raise the money, and the Centennial Commission, which recommended how to spend it.

None of this mumbo jumbo made sense to me then and it still doesn't. To put it mildly, the concept of "conflict" was hogwash.

Apparently, some of our top people in Washington don't know the difference between "conflict" and accountability. The real victim of all this silliness was the principle that President Reagan talked about and held up as a shining ideal: American volunteerism. I volunteered and they fired me. I sure hope they don't need volunteers again anytime soon.

After my firing was announced, my mother called me. She said she'd heard on the news that I got canned and wanted to ask me some questions. Mothers can get pushy with their kids. "What's going on?" she said. "I mean, is this gratitude? Is this the way the government treats you? You succeeded, and they do this. What if you had loused up!"

Frank Sinatra phoned too. He was so upset he called the President direct and asked for an explanation. He didn't get one. Mario Cuomo was so indignant he dashed off a letter to the President demanding I be reinstated. It's amazing how these Italians rally 'round you when you're in trouble.

Here's the final irony. The Monday after I received that wire, Secretary of the Treasury James Baker, standing before television cameras and the largest collection of microphones I'd ever seen in my life, presented me with a check for $24 million, representing the proceeds raised to date from the sale of Treasury coins dedicated to the restoration campaign.

Said Secretary Baker: "This effort to restore the Statue, headed so ably by Mr. Iacocca, is the classic example of American volunteerism."

At approximately one P.M. on the same day, I was half a mile away in the office of Secretary of the Interior Donald Hodel. With no cameras and no microphones, Secretary Hodel handed me a letter terminating my chairmanship of the Statue of Liberty–Ellis Island Centennial Commission.

I was the only person in the world who was present at both ceremonies, conducted within two and a half hours of each other by two members of the Cabinet of the United States of America. Had I not been there, I wouldn't have believed it, but I have the transcript of what Secretary Baker said and the letter that Secretary Hodel wrote—and I believe it. I don't understand it, but I believe it.

A *Los Angeles Times* editorial summed up the whole thing in one four-letter word: "D-U-M-B."

"What's done is done," as Shakespeare would say. My blood pressure went up for a couple of days, but it's okay now. I guess when you've been fired once, you get used to it.

The important thing was that I was determined to stay with the project to the end, politics or not. It started as a labor of love and I sure wasn't going to let it end on this kind of note. So I stuck it out, and I'm glad I did. Because the grand finale made it all worthwhile.

There were so many wonderful moments during those four years I spent raising money for the Statue that I'd like to share a few of them with you, because they got me thinking about where the country is today and where it might be headed tomorrow.

Symbols mean nothing if the values aren't there. We didn't spend millions of dollars just so that the Statue wouldn't fall into the harbor and become a hazard to navigation. We didn't fix up Ellis Island so that people would have a nice place to go on a Sunday afternoon. We did it because we wanted to restore, remember, and renew the basic values that made America great.

When my Chrysler people down in Washington first told me about the restoration project, they recommended against my getting involved. I was up to my ears at Chrysler, and they thought this job would be too time-consuming—and maybe not prestigious enough. The reason: It was to be a Department of the Interior commission, not a presidential commission.

Big deal, right? Yet it shows how even then, our government bureaucrats didn't understand the importance of the event.

Back when I was six years old and again when I was eleven, my father took me and the family in a beat-up old Ford from our home in Allentown, Pennsylvania, to see the Statue of Liberty. He walked me up all the stairs to the crown. Then he told me about how great it was to be free—and how great it was to be an American. The scene of the harbor is still vivid in my mind, but frankly I didn't know what the hell he was talking about.

Only now can I begin to understand why my father made that trip twice with the whole family in tow.

My kids have been to New York a hundred times, but only to Broadway shows, the museums, and good restaurants—and to shop. We did the town lots of times. But in all those visits I never thought once of taking either of my daughters to the Statue of Liberty. What happened in one generation? I can't figure it out myself. Why didn't I ever say to them: It's great to be free; it's great to be an American; let's go to the Statue and reminisce a little bit about who we are.

In fact, I never went back myself until 1982, when I was asked to lead the effort to restore the Statue and Ellis Island. Then I had to ask myself why I hadn't been back in more than fifty years, and why I'd never taken my kids out to Liberty Island to climb all those stairs, the way my father took me.

The answer was simple, and it was embarrassing. The Statue of Liberty didn't mean the same thing to me that it did to my father.

My parents worked hard and they built a great life in America. So my father went back to the symbol of it all, just to say thanks in his own way. Millions like him have gone back over the years, dragging kids or grandkids who sometimes didn't understand what was so special about the giant lady with the big torch or why they were climbing all those stairs that seemed to go up and up forever.

These thoughts filled my mind when, on May 18, 1982, in the splendor of the East Room of the White House, President Reagan kicked off the drive to restore the Statue of Liberty and Ellis Island, the gateway for 17 million immigrants. We guessed the job would cost us $230 million, and I didn't have the foggiest idea where to start. There wasn't so much as a dime in the kitty.

I began with a donated chair and a desk; a part-time secretary; Paul Bergmoser, whom I enticed out of retirement for a second time; and Steve Briganti, a guy who had run the United Way fund in New York. We had our ups and downs, but I must say the group got up to speed fast and became a solid, hard-hitting organization that ran for a full four years and raised $305 million. (That's $75 million more than the original goal. How's that for going over the top?)

In mapping out our strategy, we wanted to raise much of the

money from individual Americans. But we also knew that you don't come up with $305 million just by passing the hat. You have to do a little marketing.

It could only happen in America: From day one we fought the charge of overcommercializing this great symbol of our freedom. For ninety-six years she had been neglected, most notably by the press in New York City. Now these same press people became indignant that we would dare commercialize the Statue by asking businesses to contribute as well.

Ironically, in November of 1885, none other than Frédéric-Auguste Bartholdi, the Statue's young sculptor, sold to French businesses the right to use the Statue as an advertising logo in promoting their products. Poor guy, he was accused of being too commercial then!

So you can't win. Bring business into the picture and immediately you "commercialize" things. Bring government in and you "politicize" them.

To everyone's surprise, it wasn't the corporations who picked up the biggest part of the bill. It was the people. And they did it with pleasure.

At the very beginning I was worried about whether we could round up enough individual contributions. When I did a little research on the Statue's history, I found out that after the French had sent over the Statue, our government had refused to pay for a base. Some people started a fund drive, but it didn't work—until Joseph Pulitzer got involved. An immigrant from Hungary, he owned a newspaper, the *New York World*, and made the Statue a major cause there, especially among the silk-stocking crowd. He had such a hard time getting contributions that in the end he had to embarrass people into giving by putting their names in the paper if they refused.

In desperation I thought of trying the same thing myself. Can you imagine the flak I would have taken from the press for such a crass marketing ploy?

But it wasn't necessary. Once the word went out that the Lady was in trouble, a lot of patriots came rushing out of the closet.

Early on, a man walked into my office out of the blue. "Mr. Iacocca," he said, "I'm here to give you a check." It was for $1 million. I thought he was part of that old TV show *The Millionaire*, where the guy looks you up in the phone book and gives you a million bucks. He told me: "You can keep it, but only on one condition—that you agree right now never to reveal my name."

This man had come to America as a young boy with his mother. The family got rich, and he wanted to give something back to the country and at the same time honor the memory of his mother.

I had a lot of fun with that one. As I went around the country making speeches to raise money, I promised anyone who had a spare million that I would keep them anonymous too.

During the four years, I met thousands of other contributors. Maybe they gave five bucks, or ten, or a hundred, but the sentiment was always the same: It was a debt owed for a promise kept.

There was the man from Poland who sent us $2 for "this beautiful symbol" that he never expected to see himself, but at least he could dream about it. And there was the money order that arrived from a refugee camp in Thailand. Seventy-eight homeless Vietnamese had passed the hat and come up with $114.19 as "our humble share for the rehabilitation of the Statue of Liberty on the occasion of her hundredth birthday."

That one floored me. These were people who had lost everything—everything but hope. And the Lady was the symbol of that hope. They simply were pleading with us to "keep the torch lit."

And I'll never, but never, forget the saga of Mary Miller. She was just a little old lady from New Jersey who wanted to help, and so she sent $1,000. Naturally, we sent her a nice thank-you note. I guess she must have appreciated the note, because she then sent another check—this one for $50,000.

After that, I called her to say thanks again, and she said, "You guys are really nice." She was so pleased that she said she was sending along a "little envelope." When it came, it had three checks in it—one for $25,000, one for $50,000, and one for $75,000.

So much money was pouring in from her that I used to kid about it. I'd say, "I never even met Mary Miller. All I did was thank her and she sent me $201,000. Imagine if I'd sent her flowers!"

I was so grateful to her that I wanted to bring her over to visit Liberty Island with me. On two occasions when I was in New York I called her, but she was always too busy. She was eighty-seven, and yet her schedule seemed to be tighter than mine. The last time she turned me down for our date, she told me it was out of the question for her to get away that day. She was doing her spring housecleaning!

One of our most important constituencies was the school kids. Nickels and dimes streamed in from kids who baked cupcakes, sold T-shirts, and washed cars. Some of them even sent us their lunch money. I opened my mail one day and there was a letter from a six-year-old boy that said: "Dear Mr. Iacocca, here's my allowance for this week." There were two $1 bills attached. He added, "Spend it wisely!"

I also got hundreds of crayon drawings of the Statue from budding little artists and hundreds of poems in her honor from budding little Emma Lazaruses.

Everybody got into the act. One guy rode a motorized surfboard more than three thousand miles to raise money. Believe it or not, we even received $2,000 from the Hell's Angels! Underneath their leather jackets, those guys love America too.

The Pioneers of America, an organization of the Bell System, held a bake sale and in one night raised $240,000. That's a lot of cupcakes. They followed with a second one and, in a giant American bake-off, came up with $425,000 more.

I got one letter from a woman in her eighties with a check for $10,000. She added a P.S. asking for a free copy of my book. She said there was a two-month wait at the library and she never bought a hardcover book, because she didn't think it was good value.

Now there's a woman after my own heart. She'll give ten grand for the Statue, but she isn't about to go to the bookstore and spend $19.95 just to read about some Italian kid's ups and downs in the car business.

And there was eight-year-old Michael Haverly of Indianapolis who inspired everybody. In spite of a physical handicap and twelve operations, he went out and gave speeches to schools and civic groups and raised more than $6,500 all by himself.

When I took on the Liberty project, I thought I had an almost impossible job in front of me. But with millions of remarkable people like these, I never had an easier sale.

I must admit there were any number of weird moments during the long campaign. At one point we decided that we'd have the President present a medal of liberty to the ten people who best exemplified the immigrant success story, ten living Americans who were born overseas and made their mark here.

We assembled a group of prominent people to serve as judges and held a meeting at the 21 Club. The committee included such notables as Mario Cuomo, Arthur Schlesinger, Jr., Ted Kennedy, Barbara Walters, and Teddy White. Old Teddy found the meeting so boring that he excused himself to go down to the bar at 21 for a double belt of bourbon. It was about ten-thirty in the morning.

When he returned, a half hour or so later, he sat down next to me and said, "Well, where are we?"

I said, "We still haven't picked an Irishman or an Italian."

Little by little I realized that we had bitten off more than we could chew. At one point somebody said, "You can't just pick ten without picking a black." And I said, "You've got to pick at least one woman." Another said, "You've got to have a Hispanic." We had to have somebody handicapped. Then we had to have an astronaut. I thought for a while that we might have to come up with a freckle-faced Indian.

Finally we threw up our hands and agreed to choose twelve individuals because we couldn't settle on ten. We had made ourselves an equal opportunity employer.

Matters still didn't improve all that much. As the selection process wound down, outspoken Teddy White felt we had dug ourselves into a hole: "We went from ten to twelve, and so far we've got two Chinese and four Jews, and we're supposed to be talking about the great immigrant experience. We don't have a single Irishman or

Italian. If Teddy Kennedy can't come up with one lousy Irishman, and Iacocca not one Italian, we're really in a helluva mess."

And, boy, was he right. The furor still hasn't died down. You'd think out of eleven million Irish and Italian immigrants we could have found *one*.

As the big weekend drew closer, everything built up to this crescendo of liberty, liberty, liberty. Merchants began selling souvenirs and gadgets of all kinds. Bagels in the shape of little Statues were being sold as lox and liberty. One deli owner even had a big chopped-liver Statue in his window. There were Liberty shower curtains and genuine Statue of Liberty pasta earrings. Down at Battery Park you could even get your picture taken with cardboard cutouts of either Ronald Reagan or me. (I'd like to thank the 450 people who chose me, because I got a $450 royalty check from the pictures.) But I guess that's America. Everybody gets into the act.

One of my final chores was to round up a few celebrities to perform at the weekend's ceremonies. I was gratified that everyone we asked to participate—including Frank Sinatra, Willie Nelson, Peggy Fleming, Liza Minnelli—said okay, all except Bruce Springsteen. He seemed a perfect choice, since he's a Jersey kid who made all his money singing about the workingman and the work ethic. We called him and called him, and his agent kept saying "We'll get back to you."

I'm still waiting to hear from Bruce.

But even with Bruce sitting it out, we went on to have probably the finest celebration in the history of man.

I'll always remember Liberty Weekend as the longest hundred hours in my life—but also as the most joyous.

Everyone was in a party mood from the word go. And the spirit just rose higher and higher.

Since the Chrysler-Plymouth dealers had invested $5 million as the fund's original seed money, Chrysler treated them to a grand entrance. The company flew them to Paris to relive the history of the Statue. Then they boarded the *QE2* and sailed across the ocean to see the festivities. I traveled with them as far as Bermuda, where I had to get off to collect one last check at a ceremony near Hamilton.

When I arrived in the Big Apple on July 1, there were parties galore. I managed to make it to at least a dozen of them. My favorite was the one we threw for the hundreds of workers and staff who had done so much to make this weekend possible.

The parties built up the anticipation for the big moment on the night of the third. The Statue of Liberty had been dark for three years while we repaired what time and weather, and even our own neglect and indifference, had done to her. Now we were finally going to unveil Miss Liberty and find out whether the plastic surgery had been worth it.

We gathered on Governor's Island, with President Reagan and President Mitterand of France as the honored guests. It was unfortunate that it turned out to be the coldest July 3 in the history of record-keeping in New York. The wind-chill factor put the temperature well down in the twenties. Mayor Ed Koch had some clout and used it to get hold of some old army blankets that we huddled under to keep warm. I played kneesies with Ed and Matilda Cuomo—it was three to a blanket.

But what really warmed us up was the ceremony itself. I introduced President Reagan, who said a few welcoming words, turned around, and as he gazed out across New York Harbor, pressed a button that activated a laser beam. It shot a mile across the harbor and set off a light show that gradually revealed Liberty from her green toes to her mighty face. First, beams of red lit up the base of the Lady. Then they caught the pedestal. Seconds later, a blue light covered the whole monument. Finally, floodlights bathed the entire Statue in an intense white light.

Next came the dramatic finale. President Reagan threw a switch, and high above the harbor the Statue's new torch suddenly flamed into brilliance. A five-minute fireworks display lit the harbor with skyrockets of red, white, and blue, as a hundred-piece orchestra and a five hundred-voice chorus performed "America the Beautiful."

It was a breathtaking sight. When the President threw that switch, 242 million Americans were united as never before in a renewal of the spirit and the guts that built this great country. As I watched the fireworks exploding over the Statue and listened to the music, I was never more moved. If you didn't get goose bumps then, you had to be dead. And I've got to be honest: I was thinking a lot about my father.

After the evening's events wound down, it was back to reality. There was an incredible traffic jam of boats in the harbor and all of us had one hell of a time getting off the island. As was widely reported, Frank Sinatra blew his stack. I told him: "Hey, Frank, I can't get off either."

It was hours before I finally flagged down a boat, and it was two in the morning before I managed to get to sleep. Four hours later I had to head right back to where I had been to host the Tall Ships review.

Exhausted as I was that morning, the ships woke me up fast. They were a stupendous sight. Twenty-two majestic Tall Ships led the way into the harbor past the Statue of Liberty. Following behind them was the people's armada, a flotilla that included canoes, yachts, kayaks, ferryboats, naval gunboats, Chinese junks, even an aircraft carrier—something like forty thousand boats in all. I've seen gridlock before, but this was the first time I'd ever seen boatlock.

That night an awesome fireworks display, the largest in American history, erupted over upper New York Harbor. Two million dollars' worth of fireworks exploded in just thirty minutes. We needed so many of them that we had to hire three of the world's biggest fireworks manufacturers, because no one company could handle the order. But, boy, was it worth it. My ears are still ringing.

On Sunday the fifth, the partying continued in full force. I went to the Great Lawn at Central Park, where more than 800,000 people had assembled for a free concert by the New York Philharmonic. I've seen 100,000 people in a sports stadium, but I've never seen anything like 800,000 people. That's more people than live in the state of Delaware.

Right after I got there I was hustled into a little trailer, where I cooled my heels until I was introduced. Also waiting there were Police Commissioner Ben Ward and Mayor Koch. When they saw me, Koch said, "You know, this is unbelievable. There's only one day to go and there's been no crime. The muggers took a holiday. I guess they must all be patriotic."

It really was pretty incredible. The city had braced itself for the absolute worst. For one thing, this was right after our reprisal raid against Libya's Colonel Qaddafi, and the threat of terrorism was lurking over us. If a guy was wacky, what better time to try something than while we were caught up in a big celebration? So the city had imported all sorts of riot squads, mounted police, FBI men, and state militia. I guess they had everything but tanks. And yet they were all yawning by Sunday night. Nobody had robbed a bank. Nobody had mugged anyone. Nobody had even stolen a hubcap.

I said to myself: "Maybe this has been such an outpouring of sheer joy that even the bad guys have been swept up in it. If we had a celebration like this every week in New York we could solve all its crime problems."

The final big bash came on the sixth at Giants Stadium in the New Jersey Meadowlands. Fifty thousand people turned out for a three-hour blowout. I wanted to use the occasion to give the guys who actually sweated over Miss Liberty their due—the 150 artisans and workers who had used everything from computers to baking soda to make her pretty again. (The baking soda was needed to give the Lady her final facial.) So I brought all of them onto the stage and presented the superintendent with a plaque.

That was the point at which I had intended to introduce Bruce Springsteen, but he still hadn't called. Willie Nelson proved to be a great substitute.

Willie was followed by a terrific show with a cast of 15,000. It included Kenny Rogers, Elizabeth Taylor, 1,000 tap dancers, and an 850-member drill team. I thought it beat any New York Giants football game.

David Wolper, who put it all together, did an amazing job. In retrospect I wish he had skipped the two hundred Elvis Presley look-alikes. As usual, the press was merciless. They were still yapping about commercialism! But Wolper worked like a dog for over a year and never got a dime for himself.

In every respect, Liberty Weekend was one helluva party. Once the bands stopped playing and the ships sailed back home, I knew that we'd given the Lady and ourselves more than a mere face lift.

And so I'm awfully glad I devoted those four years to the fund-raising effort. At the same time, when I think about all that bureaucracy I had to mess with, I'm also glad it's over. I wouldn't do it again—at least not for another hundred years. I used to say, "Washington is a nice place to visit, but I wouldn't want to live there." Now, I'm not too sure I even want to visit it anymore.

If you have so much trouble getting your own government to stay on track for a couple of years to do something as wonderful as restoring the Statue of Liberty and celebrating freedom, how can you tackle the important things in life, like economic cooperation with the Soviet Union or an industrial or agricultural policy?

At one point during the long fund-raising effort, the ABC network decided to interview selected people, including me, on what liberty means to them.

They wired me up to the satellite, and I listened. I listened first to Anatoly Shcharansky talking from Jerusalem about his years in prison in Russia and about his newfound joy in being free. Then Bishop Desmond Tutu came on from Johannesburg, speaking about the freedom he was fighting for but didn't yet have.

I heard the ABC director whisper, "Mr. Iacocca, you're on in one minute." And suddenly it hit me: What the hell was I doing on the same program with these two men? I've never been anything but free. Freedom was my birthright, like breathing. I had nothing to compare it to. Like so many others, I've always taken liberty for granted.

I don't even remember what I said when my turn came. I just knew I didn't belong. Shcharansky and Tutu—they were the experts on liberty, not me!

Four years of hard work trying to restore liberty's most important symbol, and I still don't know what liberty is—at least not the same way they do. And remember, I had the advantage of thousands of little school kids writing in to tell me what it was in twenty-five words or less.

But I know one thing: I'm grateful for it, and I know that liberty brings with it some obligations. I know that we have it today only because others fought for it, nourished it, protected it, and then passed it on to us. That's a debt we owe. We owe it to our parents if they're alive, and to their memory if they're not.

But mostly we have an obligation to our own kids, an obligation to pass on this incredible gift to them. That's how civilization works: Whatever debt you owe to those who came before you, you pay to those who follow you. And the Lady got me thinking about what we owe.

One of the thousands of letters I received in response to my first book was from a Japanese medical professor at Dokkyo University's School of Medicine. He said:

> Since the Second World War, America has been a teacher to Japan. When I went to America two decades ago, I learned many things. Among them: Do not break a promise, respect a contract, encourage public morality, etc. And since returning to Japan, my family and I have been living by these tenets faithfully. Recently, however, I met an old American friend and found to my deep disappointment that he no longer believed in these principles. I think American society has changed in these twenty years.

That's a disturbing thing to hear, but there's truth in it. The Statue of Liberty may be one hundred years old, but we can't let our values get as weathered as she was. I realized from my ups and downs with the Lady that our priorities have gotten awfully screwed up. Things have gotten worse, not better, in the last few years, and I don't think we've hit bottom yet. I'm no longer even sure how deep bottom is.

I look around me and I see Wall Street executives being dragged away in handcuffs. I see a national deficit so high that I can't count the zeroes. I see the government paying farmers not to plant their land while the homeless go hungry on the streets. Something's rotten out there—and it's not in Denmark.

The only way to get this country back on track is to return to good old-fashioned horse sense. We've got to start with the basics: how we raise our kids, how we care for our sick and homeless, what it is each of us truly believes. One thing for sure, in these complicated times, we need some new rules down in Washington and some new umpires to enforce them.

My father told me that the best way to teach is by example. He certainly showed me what it took to be a good person and a good citizen. As the old joke has it, "No one ever said on his deathbed, 'I should have spent more time on my business.' " Throughout my life, the bottom line I've worried about most was that my kids turn out all right.

The only rock I know that stays steady, the only institution I know that works, is the family. I was brought up to believe in it—and I do. Because I think a civilized world can't remain civilized for long if its foundation is built on anything but the family. A city, state, or country can't be any more than the sum of its vital parts—millions of family units. You can't have a country or a city or a state that's worth a damn unless you govern within yourself in your day-to-day life.

It all starts at home.

HEART AND HEARTH

I
MY KIDS

Saint Hugo's Church in Bloomfield Hills, Michigan, is one of the most beautiful churches I've ever seen. Small, old, built of solid stone, it resembles a miniature cathedral set on a hill looking down on lush grounds that always remind me of an English countryside.

On the morning of Saturday, June 21, 1986, the church never looked lovelier. It was a glorious bright day; the sun was pouring in and the place was full of flowers. (I know there were a lot of them, because I got the bill.) One of my closest friends, Vic Damone, was singing, his wonderful booming voice filling the church.

Of course, I was probably a little biased. Kathi, my oldest daughter, was getting married.

I have to admit the idea had taken a while to sink in—for both of us. In the weeks before the wedding, Kathi used to kid with me: "Do you think I'm really going to go through with this? Or am I going to get cold feet?"

I tried to reassure her, but actually I was getting more nervous than she was. During the days leading up to the big moment, I played a little game with her to keep things lighthearted. I had an artist do up about a dozen placards, and on the Tuesday before the wedding I left the first one in a purple envelope on Kathi's chair at the dinner table. The card said 100 HOURS TO GO!

That was just the beginning. Every few hours, when Kathi least expected it, another brightly colored envelope would appear, each one counting down the hours, almost to the minute, when she would be stepping up to the altar. One ended up under her pillow. Another dangled from the mirror inside her car. I had her sister, Lia, hand her one at her bridal luncheon the day before the wedding. The last one I gave to her after she had put on her gown and came down the stairs. It was yellow and it said ZERO HOUR. THIS IS IT!

Kathi looked like a princess on her wedding day in a necklace and earrings I had given her as a surprise. But it was a bittersweet time; she missed her mother a lot. During the ceremony she wore her engagement ring, whose center stone was her mother's engagement diamond. And she wore the garter that Mary had worn when she married me, and brought an extra one to throw. Fortunately, a close friend of Mary's, Isla Small, had taken over a lot of the duties of mother of the bride. We even called her "HMOTB"—Honorary Mother of the Bride.

My emotion as I walked down the aisle was sheer happiness. I liked Kathi's choice of husband: Ned Hentz, an advertising copywriter who had been her classmate at Middlebury College. You can't pick your children's mates; you just hope for the best. And I think I got the best.

The ceremony was very formal, and the church was jammed to the walls. As we walked down the aisle Kathi was nervous and said to me: "Hold my hand." I was a little wobbly myself. As I looked down the long white runner to the altar, a hundred feet away, my mind was flooded with memories of the past. This was the same aisle I had walked at Lia's baptism. And for an instant I could actually hear her wailing as the priest put the salt on her lips. I could see both the girls walking down the aisle for their first Holy Communion, in their little white dresses with the big red bows. And I could faintly hear the strains of "Ave Maria" at Mary's requiem mass. I thought of the dozens of wedding ceremonies I had attended with Mary over the years. She was sentimental and always cried. Not today. Yet in spirit she was all around us. I almost felt she was smiling and waiting for us as we reached the altar.

The trumpet voluntary was ringing in my ears, and I thought simply: *This is life—the joys, the sorrows, ever-changing.* Then you pass on the baton to the next generation and, you hope, they repeat the cycle all over again.

When we finally reached the altar, I kissed Kathi and for a moment I clutched her, not wanting to let go. Then reality stepped in, in the form of six-foot four-inch Ned. And he took her from me.

I went to my seat with tears running down my cheeks, and I vaguely remember three or four people yelling at me not to trip over Kathi's long white train.

In a lifetime you get only a few moments like that, so you have to drink them in to the fullest. I had said that to Kathi before the wedding, and now I was saying it again, but mainly to myself.

A year later I had given away Lia, too, to her high school sweetheart, Jim Nagy. This time I used roses to count down the hours. Two daughters, two down, and none to go. I ran the same gamut of emotions I had with Kathi, but maybe on this day they were even a little stronger, because I knew I'd never walk down the aisle this way again. I felt a bit sad, but one look at Lia's radiant face and I snapped right out of it. For she, too, looked like a princess—and, believe it or not, I did trip over her train. If life got any sweeter than this, I couldn't stand it.

I worked all my life to make my kids happy. These moments celebrated what had to be my ultimate accomplishment: to enjoy life with my parents and then, as they grew old, to enjoy my kids and watch them start their families. I hope Kathi and Lia will be happy raising their kids. I sure was happy raising mine.

You know, of all the tough jobs I've had in my life, none has been tougher or more important to me than my job as a dad.

I've always felt that when I die, if I can say I've done well by my family—and that's all of them, my father, my mother, my sister, my daughters, and maybe even grandchildren—then I've lived a full life and a good life. What else is there? You get up in the morning, you go through the same day and night as everybody else—in Michigan, that's only eight hours of daylight in the winter, sixteen hours in the summer. You eat three meals a day if you're healthy and you're not on a diet. You probably sleep about seven or eight hours a night, and you spend the same amount of time at work, give or take an hour. So when it's all over, what can you tote up? Maybe you've truly enjoyed life. But how? You're only as good as the sharing you've done with those around you, especially your family and friends.

I wish I'd grown up in a family with ten brothers and sisters. I always liked the camaraderie of the big Italian and Irish families. Some that I knew were so big they used to call the kids by numbers. "Where's our number one boy today?" "Get cracking on the books, number four boy." And "number five girl, you help Mom with the dishes tonight." My next-door neighbors had four girls. They didn't feel the need to use the number system. Instead, they named their girls Mary, Sherry, Terry, and Kerry. Lots of confusion there. The kids were never quite sure who was getting yelled at.

I had only two, but I always counted my blessings as I thought of my many friends who never had the joy of even one.

I wanted kids badly, but I wasn't sure that Mary could have them because of her diabetes. Both the girls were born by Caesarean section and both of them began life struggling. I'll never forget Kathi's first days in the hospital.

Every morning the nurses would wheel the tiny carts down from the nursery. You'd wait in your room and listen for the sound of the little wheels coming. The third day, all the wheels rolled right by Mary's room. Mary panicked. The nurse told her she couldn't have her baby today because Kathi was very sick. Mary cried, and I must confess I did a little too.

Kathi had come down with a staph infection, the scourge of the pediatric wards. The doctors had put her in an incubator. They placed her on the critical list and didn't know if she was going to pull through. To make matters worse, it was the Fourth of July weekend. Have you ever been in a hospital on a holiday? You can get the doctor's answering service or, with a little luck, rustle up a resident or a nurse who's covering for three other nurses.

When I got to see Kathi, she was in this small incubator with intravenous tubes dangling from her little feet. She was scarlet red. I was terrified. I checked on her every couple of hours, and I lied to Mary and told her Kathi looked great and her fever was going down. After about a week she did get better and Mary was able to feed her. After another week we got to bring her home.

Then when Lia was born, she was even more fragile than Kathi. She weighed only four pounds twelve ounces, and she had to stay in an incubator until she reached five pounds and held it. Would you believe it took a full ten days to get those four ounces on her? (Today, she's on a diet and trying to lose four pounds—and is having about the same degree of difficulty. Sometimes life can be cruel.)

Maybe our daughters seemed even more precious to us because of their rough beginnings. Not only were they fortunate enough to be born, but they had also made it through those first perilous days. So I doted on them. In those days I traveled a lot, but I never came home without a toy or an animal or some kind of "goodie," as they called them. I used to go often to Cologne, Germany, and from there I would always bring back a Hummel figurine. Twenty years later, all those little knickknacks are still on their shelves; I always smile when I walk by them.

I also liked to buy them clothes. I used to carry their sizes on a small card in my wallet. Then if I found a little dress or a pair of shoes I thought looked right for them, I'd grab it. Ed Blanch, our controller at Ford, and I would often go shopping together. I remember once running out with him during lunch hour in London because Harrods was having a big sale on girls' rugby shirts. We got back a half-hour late for the afternoon meeting—and it was a very important one. Everyone there was trying to decide whether to build a new $500 million assembly plant in Valencia, Spain, or in Algeciras, next to the Rock of Gibraltar, where the land was cheaper. It was a long, tedious afternoon, but I was happy as a clam because I had gotten 40 percent off the girls' rugby shirts.

Mary and I had a very simple approach to raising our daughters, based on the common sense we had been taught by our own parents.

Unlike our parents, though, we were very conscious of the fact that our kids were growing up in fairly affluent surroundings. We had to make them understand that the values of life were not what color car do I get for my sixteenth birthday, or can we spend the Christmas holidays in Hawaii? I've often heard people say, "Hell, I got the kid everything he ever wanted. I gave him a Corvette for Christmas, and I threw a big birthday bash for him at the club. I promised him a helicopter. So why did he turn out to be a junkie?"

Life can be difficult for kids born with a gold spoon in their mouth, because they never really get to find out if they're able to work hard and make it on their own. Mary and I were determined not to fail our daughters in that way.

I'll never forget when we first moved into our house in Bloomfield Hills, and Halloween rolled around. In our old neighborhood dozens of kids used to come knocking on our door, and we expected the same here. All the candy was ready, the pumpkin was in the window, and the lights outside were turned on to welcome the trick-or-treaters. But nobody showed. We were half a mile up a winding lane, and the kids in this high-rent district didn't travel that far in the darkness. Finally, after a couple of hours, up came our one trick-or-treater. The little girl arrived in a big, long car driven by a chauffeur. I never quite got over that. Nor did I get over having to eat miniature Baby Ruth and Snickers bars for almost a year!

Years later, Andrew Brown, a friend of Kathi's, came from Chicago for her sixteenth birthday. As he drove up he saw Mary's vegetable garden and this big scarecrow right in the middle of it. The scarecrow was dressed in an old tuxedo of mine. He took one long look, shook his head, and said, "Wow, what a classy neighborhood."

Given such an environment, we had to be especially careful. Kathi and Lia are trying to forget the number of times they've heard me say "Money doesn't grow on trees." I must have said it plenty, because to this day they are really tight with a buck.

Nobody was going to wait on them hand and foot either. When the girls were little we had a housekeeper named Mrs. Shady. Lia used to call her the "sergeant" because she was so tough. Even though we had her to help out, we wouldn't let the girls just drop their stuff for someone else to pick up after them. Mary made sure they cleaned their own rooms, especially their closets, which were regular disasters. To this day Lia has on her bedroom door a brass plate that reads PRIVATE MESS, which we picked up in San Diego at a navy-surplus store.

The girls were always a little bewildered about whether we were that well off, anyway. Once, when I was a vice-president at Ford, Lia was asked in kindergarten what her father did. She answered, "I'm not sure. I think he washes cars."

My parents spent a lot of time with me, and I wanted my kids to be treated with as much love and care as I got. Well, that's a noble objective. Everyone feels that way. But to translate it into daily life, you really have to work at it.

There's always the excuse of work to get in the way of the family. I saw how some of the guys at Ford lived their lives—weekends merely meant two more days at the office. That wasn't my idea of family life. I spent all my weekends with the kids and all my vacations. Kathi was on the swim team for seven years, and I never missed a meet. Then there were tennis matches. I made all of them. And piano recitals. I made all of them too. I was always afraid that if I missed one, Kathi might finish first or finish last and I would hear about it secondhand and not be there to congratulate—or console—her.

People used to ask me: "How could somebody as busy as you go to all those swim meets and recitals?" I just put them down on my calendar as if I were seeing a supplier or a dealer that day. I'd write down: "Go to country club. Meet starts at three-thirty, ends four-thirty." And I'd zip out.

The same with Lia. I'd go to see those cute little plays, which were often deadly, but when she'd see me in the audience and give me a shy wink, I'd be happy and proud I was there.

Once I picked up Lia at Brownie camp. She was six years old and came running out to the car in her new khaki uniform with an orange bandana around her neck and a little beanie on her head. She had just made it into the Potawatami Tribe. She had hoped to join the Nava-joes, as she called them, but she was turned down. Still, she was excited, and so was I. Funny thing, I missed an important meeting that day, but for the life of me I have no recollection of what it was.

At Chrysler, I'm always talking to my senior executives, stroking them in a way. I do that with my top people. Why wouldn't I do it with my own kids?

I never sat and lectured the kids every week: "Here's the way I want you to grow up." But I did give them some simple guidelines: Whenever you have a problem, come talk; don't keep it inside you. Never lie to anybody. Never go into hock. Never borrow from a friend and forget it. Don't be a deadbeat. Don't make promises you can't keep.

My father was full of stories and parables to teach me those values. When I was young, for instance, I used to envy some of the richer kids in town. My father would say, "Be grateful for what you've got. Envy won't only turn you green, it'll kill you." To impress this on me, he told me a story which I've never forgotten.

Once there was a king who had two sons. The older boy was very much the favored one, and the younger son was always in a rage because his brother got more than he did.

One time the king said to the younger son: "I'm going to do something wonderful for you. I'm going to grant you any wish you want. You want your weight in gold? You want fine Arabian stallions? You want a harem of nubile maidens? Take your pick."

And the younger son said, "Gee, Pop, that's fantastic. Give me a day to think it over."

"Of course." Then the king added, "But since your brother is older, whatever I give you, I will have to give him twice as much."

The kid was not too happy, so he slept on it. And the next day he said to his dad, "I've made my decision. I want my right eye poked out."

Whenever I've had a tinge of envy, I remember that story. I passed it on to my daughters because I wanted them to remember the moral too.

As a parent, you have to keep in mind that you can't just do the talking. You have to listen. You have to sit down and hear what's on your kids' minds. Some people have a family council once a week, but I never got quite that formal at home. I had enough meetings at the office.

And when you're trying to motivate your kids, you have to know how hard to push them—without pushing them over the edge. Every father has great expectations for his children, but you have to be realistic. I know that in my generation there was too much pressure. My father always wanted me to go for the brass ring. He wanted me to compete and he used his brand of psychology on me, always ending up with the same message: You're never good enough, strive for perfection, and *never, never* give up. Christ, he sounded like Winston Churchill.

That was one part of my upbringing I didn't try to impose on my kids. If the girls didn't get super grades, I didn't go crazy about it. Which did not mean, of course, that I paid no attention to whether or not they cracked a book. If either of them finished below average in her class, she sure heard from me. But as long as the kids did all right, I backed off. At most, I'd do some gentle nudging. D's got me hot under the collar. C's I could live with.

For a while Lia had problems knowing what a book was. Mary started a game with her when she was in seventh or eighth grade. Mary would take a piece of loose-leaf paper and draw the calendar of school days for the month, then tape it on the blackboard in the kitchen. Every night after dinner she'd have Lia write down what she honestly thought she deserved for that school day on a rating scale of one to ten. One day Lia flunked a test and still gave herself a three. She also gave herself lots of sevens and eights. The self-rating system made her think every day about how she was doing. She couldn't kid herself, because if all month long she gave herself sevens and eights and then got a report card with a D or an F, it was trouble in River City.

Now and then I went in for incentives, offering the girls money for A's and B's, with mixed success. I tried the same thing with Kathi's fingernails. She used to bite her nails all the time, and they looked awful. And so I held nail inspection. I'd say, "Okay, those are pretty good. Those two are bad. I'll give you two bucks a nail if you can get them into shape." Maybe the money wasn't good enough, because it never did the trick. But today Kathi's nails look like a million bucks.

In some cases it was a little tricky setting a good example. Their mother and I used to tell them not to smoke. Mary, God love her, smoked all the time, and I've been known to light up a cigar. Once the kids came along, Mary did her best to cut back. She was a two-pack-a-day person, and she got down to six cigarettes a day.

Today, Kathi doesn't smoke; Lia's trying to hold it down to—you guessed it—six a day.

I did find it a little tough preaching to them on the evils of smoking while I was blowing cigar smoke in their faces. But I tried to teach them moderation in all things. Of course, you can't teach your kids that a moderate amount of lying or a moderate amount of cheating is okay. Knowing right from wrong has to be instilled in them at an early age. On some things you just can't compromise. My daughters, if nothing else, know right from wrong. (And if they don't, it's too late now anyway.)

How do you drill that into your kids over time? If I could answer that question I'd become a full-time psychiatrist and help other people. All I can do is hope that the good stuff sticks—and that the bad stuff, like smoking, doesn't.

I was always a stickler about who my daughters palled around with. From kindergarten up, there were usually a lot of kids at our house, and I thought it was important to know who they were. I didn't give them interviews, but if I thought they were weird, I wasn't afraid to say so. My kids' friends ate dinner with us often and they talked about their parents, so I knew what kind of homes they came from.

If there was one area where I was too dictatorial it was in how late they could stay out. I had a rule that wherever they were, the girls had to call in. And I was tough on their curfews. They could be out until eleven-thirty from the age of sixteen, and when they reached eighteen, they could be out until midnight.

They were never allowed to be out till two or three in the

morning unless it was for something very special. And even then they had to let me know whom they were with and where they were.

You know that old saw that comes on at night on TV stations: "It's eleven o'clock. Do you know where your children are?" I always knew where my children were. The kids used to complain: "When we're married, you'll be telling us when to be in." And you know what? Damned if they weren't right.

There's a fine line between being fair and being a dictator, and sometimes I crossed the line. Like one night a bright-eyed young chap who was just starting to show a little peach fuzz on his face called to pick up Lia. It was sleeting outside and the forecast was for two feet of snow. And here was this kid telling me he had borrowed his father's car and just got his driver's license two weeks ago. To make matters worse, he arrived carrying a thick phone book. He was only five feet tall and had to sit on the book to see over the wheel. With those particulars rattling around in my mind, I suggested it might be better if they just stayed in and watched TV. That did it. Everybody got on my ass for creating an embarrassing scene. But I didn't put it to a vote. Lia did not go out that night.

Coming in late got the girls in plenty of hot water. My policy was simple: They got grounded the same number of days as the number of minutes they were late. The record was held by their cousin Robin, who was four years older than Kathi and lived with us for a couple of years when the kids were in their teens. She was always getting into trouble. Whenever she went out, something seemed to happen. You'd tell her to be home at a certain time, and the guy bringing her home lost his way. Or there was a flat tire. Or there was a meteor shower. Finally I decided she was just conning me off my feet. After a series of minor infractions to test me, she floated in one night twenty-three minutes late and I grounded her for twenty-three days. It worked, by the way. Dr. Spock might not approve of my methods, but I was from the old school.

Tough as I was, I really melted when the children were sick. Kathi, the poor kid, had three knee operations. In the first two they took out cartilage. In the third, when she was in college, they wanted to take her kneecap out. We went all around the country together to get second and third opinions. Believe me, we found out a lot more about knees than we wanted to know.

I talked to plenty of people before the doctors took Kathi's knee apart. Finally, the surgeons operated, took out the kneecap, polished the back of it, and put it back. She suffered a lot. And it seemed as if she were in a perpetual cast. But today she's jogging and playing tennis again. She did have to give up downhill skiing, but she just switched to cross-country.

I've always been affectionate with my kids. A lot of people can't say "I love you" to anybody. Their parents never said it to them, and they don't say it to their husbands or wives. Maybe it's that puritanical holdover—that you have to be reserved; you can't wear your heart on your sleeve. I could never understand why people think they'll be faulted for doing something as natural as breathing: loving their own family.

You have to give it all you have, especially since you never know how long the good times will last. My wife died too early. She was just fifty-seven when diabetes took her in May of 1983. But at least the kids knew her when they were young. Because diabetes is so unpredictable, it can destroy you at any age. When the kids were five and ten, I used to pray that my wife hung on at least until they were ten and fifteen. And then until fifteen and twenty. Then I prayed for twenty and twenty-five, but we weren't that lucky.

I'm not complaining. There've been more hellish problems than mine. Still, I always lived with this terrible fear that if something happened to my wife, I would have to change my whole way of living. In a showdown, if I'd been faced with the prospect of the kids' being handed over to a housekeeper to raise, I know I would have quit my job.

When Mary died I went into a new phase. I had no idea how difficult it would be; it's something you have to live through. My kids were old enough to face it, even though none of us could comprehend it, but they took it badly.

Right after her death, the girls and I went away to Bermuda for a month to talk things over. We were in a state of shock, literally like zombies, as we had to deal with the fact that now there were only three of us instead of four. I had spent most of my time playing the father role, but it suddenly hit me how much a mother does. I wish more people would think about that, how much she does that you take for granted. Taking the kids to school. Waiting for them. Talking to their teachers. Shopping. Washing their clothes. Mending their socks.

It wasn't only my kids who depended on Mary. For my business trips, she had always packed my bags. I wasn't even sure where the suitcases were. She used to organize me totally. She paid all the household bills. She knew about the warranties. She kept the file on the plumbers—who were the good guys and who were the bad. Now I had to go back to square one.

The first Christmas after she died, we all went down to our apartment in Boca Raton, Florida. When we got to the door of the apartment, I said, "Who's got the keys?" We just looked at one another. Nobody had the apartment keys. Mary had always had them. In the old days we would stand there and wait for her to whip out the keys. Now none of us even knew where they were—were they on the key rack in the kitchen or upstairs in one of her boxes? The next year the same thing happened. Finally Lia went out and got a dozen keys made, but we still can't keep it straight.

I tried to ask myself: *What should I do now? How should I behave?* I didn't have many people to help me with this stuff. Sure, there were a lot of wonderful friends coming by, offering condolences and saying, What can I do? But they couldn't mother my children. I figured: *Now the kids are going to come to me with girl talk. How in the world am I going to handle that?*

Well, I did hear a lot about their heartthrobs, and I just tried my best. I couldn't put on an apron every night and say: "I'm the homemaker now. Talk to me as you would to your mother." It doesn't work that way. Things certainly were different with Mary gone, and I mean radically different.

I read books that said, when you lose a mate or a loved one, for the first six months you're in shock, and then for thirty to thirty-six months you're in a state of mourning. In other words, it takes about three years to realize that the person isn't coming back and you've got to get on with your life.

Well, whatever the statistics, that assessment proved pretty accurate in my case. It's taken me three years to get my act together, and I think I've seen the kids gradually getting out of a period of grief. Now they talk openly about their mother and all the pleasant memories they have.

I told the kids then: "You've got to throw yourself into something." Kathi was always interested in diabetes and was very active when her mother was ailing. When we set up the Iacocca Foundation to fund diabetes research, Kathi really got involved. She still is.

When Mary took ill, Lia was still in her first year of college at Hillsdale, still growing up. Her first year was tough. As a sophomore, she transferred to Oakland University near our home, so she could be close to her mother. She did much better, and all on her own. I must confess I wasn't much of a help to her with her studies. But in her senior year, three years after Mary's death, she was getting A's and B's. I was really proud of her.

Now she's graduated and has been trying to learn communications and the public relations business, her major in school.

With both girls married and working, I guess I'm headed into transition. I was almost hoping they would marry sooner rather than later, because my gnawing concern was always *What if something happens to me?* They'd be left orphans. Boy, did I worry about that. I suppose you always do after one parent dies. You suddenly realize that you're the only one between them and disaster.

Now there's almost a reverse twist. The kids wonder if they're abandoning me. While I want them to be happy and start a family, they're worried about leaving me!

I hope they don't worry as much about me as I do about them. When the kids were younger, I could never go to bed if they weren't home yet. There wasn't one night I was asleep while either of them was still out. I figure I slept over a thousand hours with one eye open.

I told myself that when they were twenty-one I'd stop worrying. I didn't. Then I said when they got married I'd stop worrying. I didn't. Now all this worrying is starting to worry me.

At Christmastime in 1986, all of us went down to Florida together. One night Kathi and Ned said they were going out to dinner and then they might take in a movie. After a while I went to bed—Kathi was out with Ned, her husband of six months, so I wasn't sleeping with one eye open, as in the old days. But when I awakened it was a quarter to two and they hadn't come in. I went back to bed, and then I woke up again at three A.M. By now Lia was awake, too, and she started pacing the floor with me. I won't say Boca Raton closes early, but at eleven o'clock they pull in the sidewalks.

We went into the kitchen and sat around wondering where they could be. Lia called The Wild Flower, a local disco joint, and it was closed. At a quarter to four I went downstairs, and instead of pacing the floor, I paced the driveway. The girls always called, and I was conditioned like Pavlov's dog to hear the bell ringing.

Kathi and Ned finally came in at five after four. They had found a disco open late in Fort Lauderdale and hadn't called because they didn't want to wake us. I was happy to see them. They were not happy to see me. Damn, I've got to stop acting like this.

Now Kathi has informed me I'm going to be a grandfather. I'm elated, but—you guessed it—I've already begun to worry about that.

My kids were always what mattered most in my life. Maybe I feel that way because the rest of my life turned out okay. But if I had conquered the world and they had turned out badly, I know I would have considered myself a flop.

As I start the twilight years of my life, I try to look back and figure out what it was all about. I'm still not sure what is meant by good fortune and success. I know fame and power are for the birds. But then life suddenly comes into focus for me. And, ah, there stand my kids. I love them.

II
GROWING OLDER

A few years ago I was invited to give a speech before the Society of Automotive Engineers at the Greenbrier resort in White Sulphur Springs, West Virginia. Like most of the other attendees, I brought along a guest. Now these business bashes can really wear you down—even I get a little bushed at times—but she was always raring to go. I told her afterward that she did okay for someone pushing eighty-one.

That's my mother for you.

I guess not too many top executives would dream of bringing along someone in her eighties to a business gathering—and they'd rather show up naked than cart along their mothers.

I don't feel that way. The Greenbrier is one of my mother's favorite places, so why shouldn't I bring her along?

A few months later I ran into Bob Galvin, the chief executive of Motorola, and he asked me how my mother was. I didn't know why he was inquiring about her until I remembered that he and his wife had really enjoyed having dinner with her at the conference. She fit in so well that now I'm beginning to wonder if people will invite me places just so I'll bring Mom.

To me, staying close to your parents is the most normal thing in the world. But plenty of other people don't quite share that view. How many times have you heard the phrase "We just drifted apart"? Geographically, maybe; if you decide to settle down near the South Pole while your parents stay put in East Rockaway, they're not going to be over for dinner too often. But even people who are a ten-minute drive away start to rationalize things, and before you know it they don't see their parents for months or even years at a stretch.

Still others never bother to see their parents at all. Their only contact may be a phone call every now and then to make sure the subscription to *Modern Maturity* magazine hasn't run out. I guess something similar happens with birds. When they leave the nest, they don't see the old man much after that.

I don't understand why it is that a mother and father bring kids into the world, try to do their best to take care of them, and then the kids give them the brushoff. For some poor people, seeing Mom and Dad is like going to the dentist for a root canal. I feel sorry for people who think like that; unfortunately, I know a lot of them.

The magic of my relationship with my mother is that with the passage of time we're as close as ever—if not more so. It was the same way with my father until he died in 1973. My mother and I have dinners and share holidays together—the obvious sort of things—but we even go off on vacations together.

Many people consider that rather unusual, if not downright bizarre. They think that when you start your own family, it's wrong to see much of your first family. You'd swear there was a law against it. Well, if you enjoy each other's company, why shouldn't you continue to see a lot of each other?

Years ago I went to California on business and took my mother and father. They had never seen California, so it was a perfect opportunity for them. We took a long drive down along the coast to Big Sur country. It was the time of the hippie movement and it was an education (hell, no, a shock!) for them to see somebody's bedraggled kids with long hair eating dog food for lunch. Along the way we bought souvenirs of handcrafted metal and turquoise, which my mother still cherishes for the memories.

While I was planning that trip it never occurred to me not to invite my parents along. After all, they were good company and fun to be around. So what if they were twenty-five years older? Gil Richards, an old friend of mine, used to say, "Hell, they're more fun than you are."

It's a big hang-up in this country that there's some sort of pecking order of age that forces you to stay in your own little group, even within your own family. As a result, so many people who have the means never dream of taking their parents with them on a vacation—or even on an afternoon drive into the country.

When I took my mother to the Statue of Liberty for the big Liberty Weekend celebration, she was hopping on and off sleek Cigarette racing boats and bounding onto yachts. I was winded keeping up with her. I can't remember her ever being happier. Now why would I deprive her of such a fabulous time just because someone might whisper: "Hey, get a load of this guy—he's with his mother"?

My own kids must have taken note of the fact that even though I was a pretty big wheel in the business world, when I knocked off and went away for my R & R, I still invited my folks. Now, when they go away on trips, they invariably ask me if I'd like to join them. I don't always go—there are obviously some trips where they don't need dear old Dad tagging along (for example, honeymoons are out)—but every now and then I take them up on the offer and we always have a good time.

Most of all, I see my mother because I enjoy her company. But I also feel a certain responsibility toward her. After all, she took good care of me for a lot of years—and I'm sure I was a little hell on wheels when I was a kid. I believe that my father did more to teach me the things you need to know as you go off into the business world. But when it comes to my values—cleanliness, godliness, avoiding bad influences—I picked up most of those from my mother.

So it's only natural now that I want to take care of her a little. She's eighty-four and she lives alone. When my father was alive, he was so dominating that she was almost totally dependent on him. She never learned to drive because of him. She was your typical raise-a-family, mind-your-own-business type of wife. Because they had such a wonderful marriage and did just about everything together, I feared she'd really come apart after he died. But she's continued to live in the same house in Pennsylvania and has done extremely well for the fifteen years he's been gone. She's really surprised me.

If you were to see my mom today, you'd be amazed. She's in absolutely fantastic shape. I'm looking forward to reaching her age if I can be in that kind of shape. I just hope the genes are the same. She remains pretty active and she hasn't lost a single step in her mind. It's incredible, but up until last year when she got pneumonia she hadn't spent a day in a hospital in her life. Even her two kids were delivered at home.

About the only health problem my mom has is a slight hearing impairment in one ear. I've been trying to persuade her to get a hearing aid installed on her glasses, but she's stubborn and refuses to consider it. I even tell her: "The President of the United States got one. Why can't you?" I've always thought I was a terrific marketing man, but I've had no luck at all selling my mother on a hearing aid. She insists they are a sure sign of old age. Imagine vanity at eighty-four!

The real marvel about her health is her teeth. Get this, she still has all her teeth, and—are you ready for this?—she's never had a cavity in her life. At the hospital they treated her like a freak because she wouldn't give the nurses her dentures. Hell, she couldn't; she didn't have any.

Even though my mother has enjoyed great health, I think she'd be better off if she weren't alone so much. She's really never been one to kick up her heels, so she leads what I consider to be too isolated a life. As a result, I feel an even deeper obligation to see her as often as I can. I make it a point to invite her to Florida whenever I go down there for a break. Not long ago I took her and two of her friends to Italy with me and we had a wonderful two weeks together.

As she's gotten older, I've fallen into a ritual of calling my mother five or six times a week. Now and then I miss a day, but not often. If I miss two consecutive days with my mother, she automatically concludes that I've contracted a dread disease or been kidnapped. And so anytime I've had a long, impossible day and haven't gotten around to phoning her, the first thing on my agenda the next morning is "Call Mom."

When I visit my mother she still treats me like a boy of fourteen. She cooks the same things for me that were my favorites in grade school. She worries about me as much as when I was growing up. I guess those instincts never change. When I tell her I'm stopping by, she goes out and shops for me and surprises me with little goodies. She still wants to please me and see me smile, even though here I am with a sixty-three-year-old frown on my face. Sometimes, when we're sitting down and having a meal, it's almost as if we've turned the time clock back to the '30s.

Besides a good home-cooked meal, I can always count on getting an ample dose of parental advice when I drop in on my mother. Her latest concern is that I'm taking on too many things, and she lives in absolute horror that in a weak moment I might drop my guard and decide to run for President. When I visited her one Saturday soon after we had announced the purchase of American Motors, she switched into her mother-knows-best style: "Why are you buying somebody else's headaches? Why do you need that? Weren't things going well enough?"

"Hey, Mom," I said, "you know what Pop used to say: 'If you stand still you'll go backwards.'"

"Yeah, yeah," she said. "But you've been saying that all your life. You could slow down a little. Remember, your father was also the one who told you to take time to smell the roses."

Without fail I hear one of her little lectures every time I visit her. She's not a nag, though. She gives me her spiel once and then we move on to other things. When I'm ready to leave, she gives me a loving goodbye and then for good measure she hits me once again with advice. That's her ritual—and so I always brace myself for it.

With every passing year, I become more convinced that you should try to do things for your parents before they get too old. If you put off sharing time with them, how in the end can you explain that to yourself? I don't care who you are—good or bad, rich or poor—as you grow older and you're sitting on a rocking chair somewhere with a sunset in front of you, reliving your life, if you've neglected your parents (or your kids), it's really got to eat you up.

No matter what you've done for yourself or for humanity, if you can't look back on having given love and attention to your own family, what have you really accomplished? Doing things for total strangers is great, but remember, giving starts at home.

I have great respect not only for my parents but for all old people. I've had this bug in me for a while now about how old people in the United States have been discarded simply because they got wrinkled. Once you reach sixty-five, you automatically get shown the door in the business world. Often, you not only get shown it, you get thrown right through it. Women have an even worse time. When a woman hits forty, she's considered over the hill by a lot of people. She's got two choices: Get a face lift or run and hide.

We've all heard the jokes by now of what defines old age. You get winded playing chess. You can't tolerate people who are intolerant. You feel like the night before and you haven't been anyplace. You sink your teeth in a steak and they stay there. The best part of the day is over when the alarm goes off. The gleam in your eye is from the sun hitting your bifocals.

Well, the jokes are turning into reality.

Once upon a time people just got old gracefully. Now, as soon as someone gets on in years, even if he's still a vibrant and healthy person, everyone says, "Gee, he used to be a great guy, but now he's seventy." They make patronizing cracks, like "Boy, he's got a lot of energy for a guy of sixty." What's that supposed to mean? I know plenty of guys who are lethargic at twenty. Besides, since I turned sixty, those comments are getting a little close to home.

Because of Hollywood and TV, this country believes in forever youthful, forever young, forever happy. It directly associates happiness with health and beauty. It's no wonder, then, that the attitude of most young people is: Sixty is getting old, seventy means you're decrepit, and eighty means forget it.

It's as if age is something evil and sinister. We use it to define a minority—even though that minority is soon to be a majority. How can we do that? Getting old has nothing to do with race or creed or religion. It's just the passage of time—and it passes for everybody. *Tempus* does *fugit*.

This notion that as people grow older they should be discarded permeates many of our institutions too. I know of a lot of companies—we've been guilty of it at Chrysler—that look at a person and conclude: "He's a good guy and he's doing a bang-up job and he always meets his budgets, but he's already fifty-three and that means he's only got about five or six good years left in him."

What makes us think he doesn't have twenty more good years left in him—or thirty?

Aside from the fact that I'm now past sixty, I guess this subject rankles me because I've always been struck by the great regard for old age in Japan. The Oriental mind says, the longer you live, the smarter you get. There's not just respect, there's reverence. Many of the people who run Japan are in their eighties. They've lived long enough and gotten wise enough at least to protect their young people from self-destruction. Nothing is new to them; they've seen it come and go. In America we wouldn't entrust someone over eighty with keeping score in a shuffleboard match.

Can we change this attitude? Well, we can at least adjust the clock. Congressman Claude Pepper has already done that. He said it was going to be against the law to make somebody retire automatically at sixty-five, and now that's come true. Unfortunately, the change in the law hasn't yet altered our attitudes. But it's a start.

As with everything else, I tend to look at this problem through business eyes. Recently I decided to keep a couple of guys at Chrysler beyond retirement age. They don't look sixty-five and they don't act it. They've got the smarts of at least a seventy-five-year-old, if not an eighty-five-year-old, and they want to stay on.

One of the adages in this country is that if you allow everybody to stay in his or her job until sixty-five or seventy-five, then the young people will be blocked from moving up the ladder. Well, if you have that problem, you'll just have to unblock some jobs. But to declare automatically that some guy is suddenly useless because he's reached a certain chronological age is nuts.

And once you do retire from your company, that shouldn't mean your work life is over. Frank Pace, who retired as the head of General Dynamics about fifteen years ago, has done an absolutely marvelous thing. He observed that all the people who were sixty-five and had been automatically retired in this country were spending most of their time playing golf or gin. The rest of the time they were sleeping. They were goofing off while there was a world out there that desperately needed help. It needed American ingenuity and know-how, and yet these folks who had all that ingenuity were applying it to what card to discard in gin rummy.

So he formed an organization called the American Association of Retired Executives, which loans out people to various countries to help them with worthy projects. They accept no salaries, only their expenses. Frank has devoted his life to the organization and has done a lot of good and set a great example for other retired executives.

There's another looming problem with the elderly that is being swept under the rug: Who takes care of them once they become ill? It's not much of a society that allows its old people to end up their lives sick and penniless and then says, "Well, that's too bad—it's just survival of the fittest. It's part of thinning out the herd." But that's exactly what we're doing.

By 1990, health care costs are expected to climb to $640 billion, a 28 percent increase over 1987. Yet even with all that money being spent, 36 million Americans have no health insurance at all.

This country as a whole doesn't seem to worry too much about the fact that people are living much longer and that someone who is eighty may be active and productive—but he also may not be able to stand up on his own. He may need to have a new kidney or he may need a new kneecap. And they don't come cheap.

We like to say we're a compassionate people. But when I look at the numbers, I begin to wonder.

According to the Surgeon General, during the next fifty years the number of people sixty-five or older will soar by 100 percent, to at least 64 million people. At the same time, the number of those twenty to sixty-four, the people who are the wage earners, will rise only 30 percent, to about 185 million.

I read another startling statistic recently that every CEO ought to tape to his desk: In 1974 the average Fortune 500 company had twelve active employees for every retiree. Now it has three.

What's more, within fifty years almost half of all health expenditures will be spent on the elderly.

It's truly a scandal if we can't guarantee all people in this country that there'll be someone to look after them when they get old. And it's true that I didn't think much about this stuff when I was fifteen, or when I was twenty-five.

But I'm not speaking for myself. Fortunately, I have the means to take care of my needs. Not everyone does. I watch my contemporaries as their parents get older and they're concerned about whether they'll be able to bear the burden of caring for them.

Well, what are you going to do when they get sick? Throw them away? Or kill them off? In some ancient tribal cultures, the very aged were pushed over a cliff. They literally went over the hill. We're doing almost the same thing, but in a slower, crueler way.

If we're going to tackle this problem effectively, the business community probably needs to jump in and become more involved. IBM took an interesting step recently when it established a free consultation-and-referral service to help its employees and retirees arrange care for elderly relatives.

I don't know how widespread the problem of the aged is. Since I've never studied it in depth, the only way I can approach it is on a personal level. One thing I know for sure is that I could never look my mother in the eye and say, "Mom, I worry about you a lot and you're home alone, and I'm concerned about your falling down the stairs. Maybe it's time to farm you out."

Why should I farm out my mother and change her life? She always says to me: "I've lived in this house all my life. It's my house, and I want to stay in it."

I hope I never have to make such a decision. I've had close friends agonize over that situation. Many times there's just no other option, and then I spend a lot of time reassuring them that they did the right thing.

Loneliness as one grows older is not something that applies only to our parents. I started to think about it myself when Mary passed away five years ago. The first six months after she died, I couldn't quite fathom what had happened. Then I realized that life had really changed.

As a result, I had a very tough time of it in late 1983 and 1984. Luckily, things were going a lot better at work. And I was so concerned about my kids that I didn't even think too much about myself at the outset.

But one day I came home and I realized that the kids were leading their own lives and they weren't at home anymore. And I found myself all alone and tired as hell. I had someone to come in and do the laundry and take care of the house, but I was eating alone almost every night. Eating alone was bad enough, but living alone really scared me. Dying alone scared me even more.

Since I come from a strong family, being alone was abnormal to me. I don't want to offend all the bachelors of the world, or all the people who never got married by choice, or who don't feel you need to marry in order to have a fulfilling life. But I really believe people were made to be together and live together.

Sure, it's very peaceful living by yourself, but it's downright lonely. Some old people say, "Hey, I've got two dogs and two cats and I love them, and nothing else matters to me." Fine. Talk to the animals. I like to talk to people.

I knew for sure that at my age and with my job, I didn't want to be out courting again. I couldn't quite see myself at a singles bar or enrolling in a video dating service.

But it's instinctive to start looking for someone to talk to and share things with. Life is full of peaks and valleys, but if you can't share your joys and sorrows with someone, then you've got yourself a pretty empty life.

So I began to entertain the idea of a second marriage. What's tough about second marriages, though, is that you find yourself getting a little selfish. I knew that because of my age I wasn't going to have any more children. Therefore you marry mainly for companionship. You hope, of course, that you're doing it out of true love, but it's not the love of starting a family and keeping the name in circulation—that's why it's a little selfish.

While I was running around raising money for the Statue of Liberty, I met Peggy Johnson and we began seeing a lot of each other. We enjoyed each other's company, and we had a lot of laughs together. She was young, she was pretty, she loved to cook, and she was a whiz at crossword puzzles. One thing led to another, and we decided to get married. But I have to admit that I didn't know for sure if I was ready to get married just yet. I wasn't sure in my heart I was done mourning my first wife.

On the one hand, the practical side of me kept whispering in my ear: *This living alone business is for the birds. I've got to have a companion who makes me laugh and who I like to be around.*

Then there was the side of me whispering in the other ear: *I don't know. I'm still going to have to spend a lot of time with my mother and my kids, and my job is still a killer, and my traveling schedule is impossible. Why not wait before I settle down again?*

In short, I was completely screwed up.

I probably should have waited a little longer than I did. But I figured I had to get on with it, and so Peggy and I got engaged in January of 1985. And of course, one of those voices was right: It was a little too soon. As a result, we didn't have a very good engagement. Early on, Peggy and I were arguing about whether I really (and wholeheartedly) wanted to get married or not. If you decide to marry someone you've known for a couple of years, you probably shouldn't have to be engaged fourteen or fifteen months, should you? Well, that's how long we were engaged. If nothing else, maybe I set a record for the longest engagement by a sixty-year-old.

But there were some sticking points we had to iron out. For one thing, there was the matter of children. I felt pretty strongly that it wouldn't be right to have kids at my age, and so I told her early on that if she married me she would foreclose her opportunity to have children. I even suggested that maybe that was too big a compromise for her. After all, I had already raised a family. She never had. But she thought about it long and hard and felt she could cope with it.

Then there was no escaping the fact that Peggy was a generation younger than I was. At that time she was just shy of thirty-six. My daughters, among ten thousand others, wondered why I was going to marry someone so young.

There were some people who wrote to me: "Don't let the age gap worry you. I'm eighty-two and my wife's thirty. If you're really in love and truly respect somebody, age makes no difference. Go for it." The majority, mostly women in their forties and fifties, told me to pick on somebody my own age.

Well, I don't think even Dear Abby can give you a rule about whether or not you can bridge a generation gap. There are plenty of pros and cons. But I thought the age issue was a nonissue. Peggy was so strong and outgoing and energetic that she made me feel young. And I'm one of those poor creatures who tend to mirror the moods of the people around me. I like to think that, mentally, I'm thirty-five. I work like a thirty-five-year-old. (Unfortunately, I put on weight like a sixty-five-year-old.) So I hated to believe that there's only one right age span for two people to make a go of it.

There was one matter, however, that turned out to be a real issue. When you're affluent and have children and you're marrying a person a generation away, a common practice today is to draft a prenuptial agreement. And so I did. In retrospect I can see that it's important. At least my lawyers thought so. But it's not too romantic. It doesn't quite match a bouquet of roses or a five-carat diamond—I'll tell you that much. Almost the instant the subject is broached, it has a way of changing the complexion of a relationship.

So, all in all, Peggy and I carried a lot of excess baggage into the marriage. Asking a woman a generation younger to have no children and to sign a prenuptial agreement as well—that's a lot to expect. If you look at all the marriages that don't make it, the experts say children and money, in that order, are the root of the problem. Those were exactly the two things we were talking about.

Some people also say that you shouldn't get married unless you're head over heels in love, the way you were when you courted your first girlfriend. They say if you're not walking on air, don't get married. I don't know. It's been so long, I'm not sure what walking on air means anymore.

The way I looked at it, I wanted someone I liked being around and whom I respected. I liked being around Peggy. I wanted someone who was a good-hearted, outgoing, friendly person. She was all of those. The walking-on-air business I could skip.

The real truth is, you're never sure. That's life. There are no guarantees that two will become one and you'll live happily ever after.

Nonetheless, in spite of all the dire warnings about the odds, it seemed that we had hacked out our differences, and so we went ahead and got married in May 1986. We had a small private ceremony in a chapel at Saint Patrick's Cathedral and went off on a one-week honeymoon to Palm Springs.

All this took place three years after Mary died. And yet Peggy and I had a very stormy first six months of marriage. No matter how hard we tried, we couldn't seem to get down to fundamentals. And I accept at least half the responsibility for the storm.

I don't know whether it was my fault for not insisting that we wait another year, but that was certainly a big part of our difficulties. Right away I started to second-guess myself. What had been the hurry? I asked. I'm independent by nature. Why hadn't I just waited until I was positive I was ready? I had already set a record for an engagement, why hadn't I really made sure nobody would ever break it?

It became very obvious that Peggy hadn't been fully prepared either. The most obvious way it showed was that she didn't want to move to Detroit. She was dead-set on staying in New York. You might wonder how on earth we could have gotten married if we hadn't reached the conclusion that we would be living where I worked. Well, I thought we *had* reached that conclusion. I assumed it. I even bought a new house in Detroit, I assumed it so much.

But whatever her reasons—and I don't think she knows what

they were to this day—she didn't want to move there. I know Detroit can be bad in winter, but not that bad.

Her refusal placed us, as you might expect, in a pretty ridiculous situation. I again was spending a lot of time alone in Detroit—and now I was married. Not only was I embarrassed, I was getting mad. I was also running up a hell of a phone bill.

Moving was our basic problem. But Peggy also felt that I lived too much in the past. She even got to the point where she thought that whenever I was pensive I was thinking about the past. Well, I was. I couldn't very well get a lobotomy and have my brain taken out.

Yet eventually, anything from the past became such a touchy issue that I almost had to make believe it had never happened. I found out that just wishing away pleasant memories was impossible. Too tough on the heart, and too tough on the nervous system.

There's also no question that it was hard for Peggy to live with all the celebrity stuff. I don't think she felt she had any identity while she was with me. And I respect her for that. How would you like it if you went out and everyone you ran into ignored you and fawned all over your mate? Who wouldn't feel like a fifth wheel—if not a sixth? That sort of hoopla can get boring pretty quick. I didn't like it either, but there wasn't much I could do about it.

Naturally, my kids didn't accept Peggy as a mother, especially since she was only ten or fifteen years older than they were. And that created a bit of a problem. Once you've raised children with their mother, it's just too difficult, if not impossible, for them ever to accept another woman in her place.

So the strains just built and built, and I began to think that maybe I could never have a successful second marriage.

Frustration was beginning to set in. I gave the situation about three months of hard thought. And I laid out a few things on paper and told Peggy that, as a minimum, this was the way it had to be, especially the part about where we were going to live. But she still wasn't ready to accept any conditions.

By December I decided that push had come to shove. We had a lot more talks, and finally I didn't know what else to do. We couldn't pick a spot halfway between New York and Detroit and live there.

We couldn't say, "Okay, we'll compromise and live in Cleveland." And so I told her: Move or else. She didn't move, and so I filed for divorce.

That was a tough thing for me to do. I always feel as if I'm in command of myself and this was an admission that I hadn't thought something out clearly enough and had made a terrible mistake. If I'm so sure of myself and so decisive a person, how could I want a divorce eight months after I got married? Well, matters of the heart don't respond very well to statistical analysis. Looking back now, I wish I had waited with Peggy another six months to a year and had not been too busy to work out all these real, live, everyday problems ahead of time.

At the time I filed, though, I thought: *Maybe I have to learn to live with the hand that was dealt me, and that's to be alone for the rest of my life.*

I'll tell you one thing. The divorce announcement blew my mother's mind. She's from the old school, which says that divorce is unthinkable. Right off the bat, she told me that I hadn't given the marriage enough time. "You work at it," she said to me, "and then you work at it some more."

I said, "Hey, Mom, that's fine advice. But if we're in two different cities, how do I do that?"

I filed the papers in December of 1986. And I guess it was a test of my love for Peggy, because I started to suffer. Really suffer. The divorce papers also had a jarring effect on her. She said: "This is getting heavy. Let's talk it over." She also agreed that maybe we should see a counselor and work through our problems.

And so in the spring we did just that. I've got to tell you, that was one helluva big decision for me. Because of the way I'm built, I'm used to solving my own problems. But I wanted to make sure I left no stone unturned. I felt counseling might do us both some good. So we decided we should make an honest effort at reconciling.

I've always believed that during the first couple of years of marriage you may have a lot of spats, but you get stronger from them. The key is that you're willing to hash out the problems and that you have basic respect, admiration, and love for the other person.

My kids accepted my new situation. They assured me: "Look, anyone who makes you happy makes us happy. If Peggy makes you laugh and enjoy life, that's all that counts."

Well, I wish this personal episode of mine had a happy ending. But it was not to be. We didn't make it. So, on November 19, 1987, at 1:43 P.M., our marriage was dissolved in the Oakland County Circuit Court. And with one rap of the gavel, sadly, it was over.

With all the professional counseling and all the amateur advice from well-meaning friends and all the long talks with Peggy, I failed. I think we both genuinely tried to resolve our differences, but after six months of trying, it didn't work. The divorce papers sounded so cold. The marriage "is irretrievably broken down because of irreconcilable differences." No time to crack a smile or even blow a kiss to what might have been. You know, I still had a lot of love for that girl.

One thing that's clear to me now is that there's no reliable road map the second time around. You really have to buckle up and drive by the seat of your pants. Well, I guess I ran right off the road—and I wasn't even wearing my seat belt.

III
FAME AND FAITH

In the last couple of years life has been chock-full of opportunities. Just the other day I heard from Everett's Pizza Parlor in Pocatello, Idaho. The owners wondered if I could knock off and come to Pocatello to help them build up their business. They said I'd be well rewarded. I'd get to become an honorary citizen of Pocatello—and I could have a large pizza with extra pepperoni on the house.

The next day I got an invitation from the Hog Callers of America. Could I possibly drop by for a visit? I had to scratch my head: As far as I could remember, I've never called a hog in my life.

Ever since I got roped into doing Chrysler commercials and my face started turning up during breaks in *Laverne and Shirley* reruns, I've had to spend a lot of time politely declining invitations. I thought I'd heard from just about everyone—until I went out and wrote a book. Suddenly the whole world knows me, and without my American Express card.

People actually seem to think that I'm a miracle worker. Believe me, their expectations are way out of line. If I were a miracle worker, you'd never see a lousy ten-day report on our auto sales.

I have people writing to me wanting to know how to get their kids off crack. Well, I never had a kid on crack. They want to know what's the best way to raise their cat. I've got two dogs.

Sometimes the questions are really far afield. Not long ago I was in Miami and some reporters asked me what we should do about Nicaragua. I was looking at my watch because I was due on the *Miami Vice* set. You can see where my priorities lie. Then, a week later, I was in Atlanta and they wanted to know how I would handle Qaddafi. I thought: *Hell, I don't even know how to handle the UAW.*

The fact is, the myth about me has gone so far beyond the reality that it's ridiculous. I'd really love to meet the guy I'm supposed to be. I'd hire him in a second.

And I'd have to be pretty dumb to take this sort of buildup too seriously. If you read all your press clippings and took them to heart, you'd be too scared to get out of bed. The way I look at notoriety, you have to keep your feet firmly on the ground or you'll find yourself nothing but a weed tumbling in the wind.

As a youngster I wanted to be voted the kid most likely to succeed, but this popularity thing is now getting a little out of hand. I'm serious. It's still a puzzle to me why my book sold the way it did. If it had come and gone in three months, and I were left with a leatherbound copy to give my kids, I'd have been reasonably happy. My publisher said 100,000 copies would make *The New York Times* Best Sellers list, but 2.6 million hardcover copies seemed a bit much. I still can't define the phenomenon. In fact, it flabbergasts me.

When the mail first began to come in, I got preoccupied with geography. I was stunned by the number of letters I got. Michigan mail I could understand. Even the Pennsylvania stuff, with its local-boy-makes-good flavor, I could understand. But when it started pouring in from East Africa, India, Greenland, western Australia, Poland, and China I was mystified. I got a letter from a young man in Changzhou who wanted to know: "What first-rate bean counter mean?" A retired worker from Luleå, Sweden, told me that my book was hot high up in northern Sweden, not far from the Arctic Circle. It seems they have plenty of time to read during the twenty-four-hour nights up there. I thought my book might appeal to the sports-car crowd, but hardly to the reindeer-sled set.

The book even had an impact in Red China. I know that because Deng Xiaoping's son sent me two books and asked for autographs for him and his dad. Apparently I was big in China because a lot of Chinese felt I had an Oriental soul. They told me they identified with a story about humble beginnings, love of family, adversity, and especially patience and perseverance.

I found out that Ferdinand Marcos read the book too. He said he learned a lot. He couldn't have learned too much—he got thrown out. Maybe he should have read it a little earlier.

The best thing about the first book was that it enabled me to make thousands of pen pals all over the world. There's a Wonder bread commercial on television that uses the tag line "Just a little slice of America." That phrase pretty much describes my mail. Every day, along with the avalanche of documents, memos, and letters about Chrysler business that cross my desk, I get hundreds of letters from people who simply have something to say and want to say it to me.

No two of these letters are alike, although many share the same opening lines: "Dear Lee: You don't know me but I feel like I know you after reading about you and seeing you on television," or "Dear Lee: I bet this is the strangest letter you've ever received, but I simply had to write," or "Dear Lee: I've been thinking about writing you for some time, but my [husband/wife/son/daughter/niece/nephew/cousin/godparent/or parole officer] said not to bother because it'd be a waste of time and that you'd never answer me," or "Dear Lee: I've always been told if you want to get something done you should go to the top, so here I am. . . ."

After openings like that, anything goes—and I mean anything!

My favorite letters come from very young people or very old people. Neither group minces words. They get to the point fast and really tell it like it is.

The kids ask questions about my favorite subject in school, what influenced me most in my life, and what I like to eat for breakfast. The senior citizens tell me what I should have studied in school, what influence in life I should have avoided, and what I should eat for breakfast.

I have to give people a lot of credit for their resourcefulness in finding some common ground between us. I've heard from thousands of people who claim to have played some part in my life but with whom I'd lost contact over the years—school chums, teachers, former neighbors, girlfriends. I've received letters from people reminding me that we had dinner together at so-and-so's house in Allentown in 1941—and mentioning that we had Yankee pot roast, sweet potatoes, a tossed salad, and Jell-O. It's amazing how many people grew up with me or dated my sister Delma or ate Yocco's hot dogs or who drive only Chrysler products (thank God for those folks). From the flood of letters I've received from people who say they were in my class in high school, I know that the class of '42 at Allentown High had to be the largest graduating class in U.S. history!

Whatever excuse people use, I read every letter addressed to me—and respond to everyone I can. So many of them amuse and touch me that I can hardly wait to pick up the next batch.

There was the Canadian woman who informed me she had named her newborn daughter (seven pounds, two ounces) Erika Lee. She sent along a baby picture too. Luckily for the kid (and me!), I didn't notice any resemblance.

I was really happy to hear from a student at the University of Denver who said he paid for his education solely from the profit he made from Chrysler stock.

And I'm always hearing from CEOs on one matter or another. The oddest one, I suppose, was the chief executive of the ministry at San Quentin prison, who wrote that "you offer a professional business and management style that has application for us. Thanks for the tips." Wait till my guys hear that I run Chrysler like a jailhouse.

Another gratifying thing about my mail is that I get a steady diet of sound professional advice free of charge. For instance, I recently received a letter from an eighty-four-year-old barber with two pages of advice on how to part my hair. By studying pictures of me that captured every angle of my head (except from the inside), he scrupulously analyzed the shape of my skull and the thickness of my hair (what's left of it). His conclusion: No doubt about it, I had to start parting on the right side.

I'm a pretty cautious guy, so I've decided to get a second opinion.

Since my big salary gets splashed across the newspapers every year, I also get plenty of mail offering me the opportunity to get in "on the ground floor" of the "chance of a lifetime."

By now, I've been extended the opportunity to sponsor country-and-western bands, swing bands, big bands, high school bands, jazz bands, and religious bands—every one of which, I'm assured, will make the world forget the Rolling Stones ever existed.

Not long ago I even got a super invitation to sink $350,000 into a snail farm. In a mere five years, I was told, it would breed more than 35 million snails. Now I love escargots, but I don't have the faintest idea if there's a market for 35 million snails.

Since I'm in a business whose lifeblood is new products, many people also write to me about inventions for which they could use a little seed money. A chap from Oklahoma thought I was just the man to sponsor his "aboveground tornado shelter." A young Maryland girl sent me word of the "One-Second Paw Cleaner." In great detail, with drawings attached, she described a sponge-filled box with four holes in its top and a roll of paper towels attached to the side. When a muddy-pawed pet entered the house, it was directed to step into the holes and then onto a sheet of paper towel to dry its now-clean paws. After some careful review of the size of the dirty-paw market, including the need in my own house, I elected to pass on that one too.

One of my most devoted constituencies is the religious crowd. A while back, for instance, a gentleman from Ohio asked for $5 billion to help build the Kingdom of God. I wondered if he'd met the gentleman from Texas who called to request a LeBaron convertible to be used as "God's Pace Car in Texas." The Texas caller later wrote to say that it would probably be inappropriate for God to ride in a hot-red convertible, so could he please change his request to a black 1988 Chrysler New Yorker?

On a really good day, somebody writes offering *me* a few bucks, for a change. A few months ago I received a letter from a German woman named Carola Gastern, who said she was getting on in years and had amassed a considerable fortune. She was so impressed by my book that she wanted to leave her entire estate to me so "you would be able to attend to your literary work more intensively."

I was quite taken by the letter, but I wrote back to assure her that I wasn't yet down to the small change. In fact, I was pretty well off. I told her she might do better leaving her fortune to an organization that could further good relations between our two countries.

Only after a bit more investigation did I discover that Carola Gastern didn't really exist. It turned out that she was actually a German male high-school teacher named Winfried Bornemann, who writes satirical books of correspondence with prominent people and government leaders. He writes nonsense letters under various pseudonyms and then publishes the answers in his books. It was just one more racket, and I was the latest guy to fall for it.

Not all the mail that I receive is simple to deal with. Life isn't easy and some people must be at the end of their rope when they decide to write me.

The letters from families facing personal tragedy always touch me the most.

Recently one woman wrote: "My father-in-law, aged eighty-two, lost his vision and suffers from congestive heart failure. Each day, we read a chapter from your book. It is one of the few experiences that gives him pleasure."

Then there was the letter from the man whose home was the Federal Correctional Institute in Lexington, Kentucky. Cocaine put him there. He wrote: "If you could turn one of the nation's largest corporations around, I'm sure I can overcome this one lousy experience in my life."

Some of the letters break my heart.

I heard from a man from Southfield, Michigan, who said: "I didn't just read the book as most people normally would. I read it out loud, every word, to my daughter. She is hospitalized for life, born with multiple major seizure disorders that prevent her from having much comfort at any given time. . . . Katie won't live a very long life, and most of it will be in heartbreaking discomfort, but at least she will have had some relaxing moments and a chance to know about you."

How could someone not be moved and humbled by letters like that?

At the end of my last book, I asked for donations to restore the Statue of Liberty. Lots of money poured in. But I've also gotten plenty of donations I never expected.

People have knitted me scarves, hats, sweaters, afghans. One man crocheted a ladies' shawl and sent it to me to give to Peggy as a gift of reconciliation.

I've gotten a Hedstrom child's rocking chair, a Statue of Liberty costume with a crown (no, I haven't worn it yet), an all-weather raincoat, a handmade Christmas stocking in the shape of a business suit, a box of Jockey underwear (yes, I have worn those), a six-foot Norfolk pine, and a pair of snowshoes.

I've gotten hooked rugs, stained glass, and paintings galore. I've also gotten quite a few sketches and prints of my face. By now there have been about a hundred interpretations of what I look like, most of them better than what I really do look like. Every day, the mail bulges with cuff links, ties, all manner of hats—fishing caps, bowlers, hard hats. I've received enough cigars from around the world to meet the needs of a major league smoking club.

Then there's the symbol fraternity. If there's just one more person who's found a new way to do a Chrysler pentastar—I've already gotten replicas made out of pretzels and chocolate—I'm going to go crazy!

Food is another favorite gift. A New York psychiatrist even sent me a box of two dozen chocolates with my face imprinted on each one. Have you ever tried biting into your own face? Talk about creepy. A psychiatrist ought to know better.

Once you've hit it big, it seems that all kinds of distinguished organizations and societies want to show due respect by bestowing honors and awards on you.

Well, I'm happy to say some real honors have come my way.

Members of the Caricature Society of America were polled on who had the best features to sketch. The winners were Diana Ross's eyebrows, Beverly Sills's lips, Jane Fonda's chin, Linda Evans's hair, and Lee Iacocca's nose. Here I've got these big brown sexy Italian eyes, and they've got to pick on my Roman nose.

I also came across a survey in *People* magazine that asked readers whom they would choose if they could borrow anybody else's brain for twenty-four hours. Albert Einstein led the list, followed by JFK, Ronald Reagan, me, God, Thomas Edison, the person's mother, and Mozart. I could use the day's rest, but believe me it wouldn't be all that interesting. Certainly not as interesting as God!

Adolfo, when it introduced its new Adolfo for Men fragrance, asked guests to name the business executive they thought most deserving of a lifetime supply of the scent. I was the lucky recipient. I don't know if that meant they liked me or they didn't like the way I smell.

Allen Paulson, the head of Gulfstream, has even named a racehorse after me. If everything goes as planned, it'll be running in the Kentucky Derby in another year or so. I told him that I was honored, as long as the horse didn't turn out to be a nag. He felt there was no way that could happen, because he paid $750,000 for the stud fee. Can you imagine getting paid that kind of money for work like that? Seattle Slew is the father, and a mare out of Native Dancer is the mother. Not too shabby parents. Next year, my friends will be able to go to the Derby and root like crazy for me—or hope I break a leg.

There's another wonderful group of people who own lodges, spas, or fat farms, and would like me to be their guest for a couple of weeks. Obviously, my jolly cheeks are starting to show. Then there are the hundreds of people who claim they cook great Italian food and want me to drop by their house for dinner. I was tempted to take up a woman from North Dakota who invited me over to sample her mother-in-law's meatball soup.

I also get some awfully tantalizing invitations. A guy from the kibbutz Palmah Tzova in Israel invited me to spend a week with him. The Gatlin Brothers Band asked me to drop by and play a little golf with them. For a roundup of "fantasies of the famous" that appeared in *Reader's Digest*, I mentioned that someday I'd like to lead a swing band and play the tenor sax. Sure enough, I got a letter from the Hot Tomatoes Dance Orchestra in Denver, extending me an invitation to lead the band.

One of these days I'm going to pick six or seven of these invitations at random and just hop on a bus and visit. What a wonderful way to meet some nice people.

After my wife died, I also got a ton of letters from women. A number of them wrote to me and said, "I spent the night in bed with you and you were fantastic." I'm glad somebody had a good time.

When the news got out about my divorce, the letters started up again. I received one recently in which a woman wrote: "I like good music, good books, good food and I'm forty-eight, attractive, and divorced. How's that sound?" Another one began: "I think I have some good news that will cheer you up. You see, I have my mother and my special aunt. They are both very interested in taking care of you. Although we are not Italian (we are Greek), they are very good cooks and would love to learn to cook Italian the way you like it."

One young lady dashed off a letter because she was sure that I had been driving a red Chrysler convertible on the Florida Turnpike on December 3, 1986. "If it was indeed you," she said, "I am the gal in the gray Cutlass you played tag with for about an hour until I had to get off at the Delray Beach exit." Alas, it must have been my double.

Probably the most creative letter I got was this one:

> I can't Dodge the fact that my skin is no longer as smooth as Corinthian leather and my chassis is not Imperial, but I come from super stock and with a great warranty. Don't panic, I'm not a New Yorker, and children—well, I'm LeBaron. Unfortunately, I'm a Charger when I should pay cash. Maybe when the moon and the Sundance on the Horizon we could go to the Lido for some turbot and a magnum of Lancers. Seriously, if you're interested in a slightly used 40-year-old (twice now I've never made it to the 5/50, much less the 7/70), I'm available. I'm not a Barracuda, just out for dinner and conversation. So, Mopar

to ya, be Valiant and remember the Challenger. Call me right now. I'll be in a Fury if you don't.

A number of these female admirers send along their pictures—to titillate me, I guess. Usually they're snapshots, though a few are professional 8-by-10 glossies. A Korean woman really intent on titillating me sent a photograph of herself topless. She made me blush. Even my secretary blushed.

Another nice thing about being a big wheel is that you get phone calls from the rich and famous. Last year I got a call from Paul Newman. I was really flattered that my favorite actor wanted to talk to me. It turned out that he and Tom Cruise were looking for $7 million to sponsor their racing efforts. I thought that was a little steep but offered to negotiate if they would do a couple of commercials for us.

Paul said that was out of the question, because he felt a celebrity was over the hill when he stooped to doing commercials.

"Paul," I said, "how about me and Bill Cosby?"

We didn't make a deal, but he's still my favorite actor.

My problem now is that far too many people call me—with far too many requests. All of them, I'm assured, are for the good of humanity, and I'll bet a lot of them are. But it's pretty hard to be in a thousand places at one time—even for someone who's supposed to be larger than life.

Ever since I got done raising money for the restoration of the Statue of Liberty, I've been invited to "pick up the torch" for similar efforts across the country. Judging by the mail, there must be literally hundreds of statues, museums, bridges, parks, and parking lots in need of repair.

Some people aren't even sure why they want me wherever it is they want me. One man actually wrote, "I don't know what it is, it's abstract, but I read your book and I feel good about it. I can't tell you what I feel good about, but I do feel good. So I'd like to get together with you."

Every now and then, I get an invitation that sounds like fun.

Then I make the mistake of accepting it. Before I know it, I end up in hot water.

A couple of years ago, Michael Talbott, the actor who plays the rumpled Detective Stanley Switek on *Miami Vice*, wrote to me for an autographed book. I sent him one and he wrote back to invite me to visit the show. I didn't watch the series myself, but I did know that probably 80 percent of the country's females were hooked on Don Johnson.

As it happened, we had a Chrysler dealer meeting scheduled for Miami in early 1986. When I arrived I had some free time and went over to the waterfront, where the show was being shot.

Next thing I knew, I was having lunch with the cast. Then they suggested a walk-on. Before I knew it, they handed me a script. Next, they gave me my own "star" dressing room, and wardrobe dropped by to make me look a little more like a slob.

The plot was super-simple. There was a big drug bust, and I was to play Park Commissioner Lido. The script described him as a "silver-haired, self-possessed, no-bullshit administrator type." In the scene, I was supposed to tip off Don Johnson and the other guys about when the drug boats would come in.

I still remember my last line: "If you guys need some help, I know how to handle a gun."

I did okay on the first take, but on the second take I started to foul up my lines. Finally I got so confused that I didn't know what the hell I was doing. I was supposed to have those ten lines memorized, but these people forgot I was weaned on cue cards. By now this little visit had turned into a six-hour shoot.

Once I left the set, I figured that was the end of it. But the next day, the publicity department went into action. There I was on the front page of *USA Today* in my Hawaiian sport shirt, trying to look cool.

The mail came quick and to the point: "If I were you, I'd stick to commercials." Or: "If your schedule is so full, why do you have time for drivel like that and you can't find time to attend the Campfire Girls bonfire?"

What you learn quickly is that when you're in the glare of publicity you've got to be careful. You shouldn't do things just for the hell of it. *Miami Vice* wanted a little publicity, and the world assumed that so did I. Some of the stories that followed said: Ah, this only proves he is a supreme egotist. Or that I was bucking for a movie role—and what better way to get to the presidency than through the movies?

The simple truth is, I went on that show as a lark. It was a nice change of pace. Now and then you've got to knock off a day and do something different in your life, just to keep from getting stale. Little did I know.

These days it seems that I even have to watch what I wear. I was doing a Chrysler truck commercial outdoors and it got cold. I didn't have a coat, so I borrowed the director's raincoat. Sure enough, the wind blew the coat open, revealing the Burberry lining. Next thing I knew, the head of Aquascutum was calling. He suggested I was totally out of style and sent me two of *his* raincoats.

When you go on the tube, people really do notice everything, and so you can't make a single mistake. Unfortunately, we recently made a whopper. I did a Jeep commercial, which unfolds with a shot of the American flag in the background and then switches to the Jeep plant, where I'm shown merrily riding along in a Yamaha golf cart. After it aired, the public flooded me with mail. Just what a Japan-basher needs!

The real problem with these commercials is that every time I go out, there's always someone who sees me for the first time in person and says, "My, you're so much taller than you look on TV." I'm told that the TV tube does tend to shrink you and flatten you out a little. But I don't believe that for a minute. I'm convinced that Ron DeLuca, who has produced all of my commercials, is doing it on purpose. He's five foot two and I know he's getting even for all the little people in the world.

By now, the myth of the comeback of the century, or of the comeback kid, has also spawned all kinds of similar stories that use me as a role model. Some of the comparisons get a little far-fetched. I can't tell you how many profiles I've read about the guy who's the

Iacocca of the sardine business or the Iacocca of the shellac industry. I felt the limit had been reached when I saw an article in *Automotive News* about Chief Phillip Martin, who was described as the "Lee Iacocca of the Choctaw Indian reservation." But then a front-page article appeared in *The Wall Street Journal* touting Gorbachev as the Iacocca of the Soviet Union. I don't know if it was meant as a compliment or an insult.

One of the crazy things that has happened recently is that a number of people, even though they like what I stand for, are trying to change me. In the last couple of years I've received quite a few letters that say: "You've got a lot of clout with college kids: Make sure you use it right. Watch what you say, because every word counts."

Well, I'm not yet part of the State Department where I've got to worry about my adjectives. I've actually started to feel a little self-conscious. Even my own guys want to tone me down. I've had to tell my speechwriters when they give me a draft of a speech that it's not strong enough for me. But they tell me: "You have to pull in your horns. You have to be more of a statesman. You've arrived. Start acting a little more like Churchill—and a little less like Harry Truman."

Arrived where? If I begin talking like I don't talk, nobody will understand me. Some of these speeches get so diluted, neutralized, committeeized, and homogenized that I don't recognize my own ideas.

There's nothing wrong with a few refinements, though, and I have altered some of my ways. I'm not going to use as much profanity in the future, because after the first book everybody jumped on me for the curse words. A typical example was the Arkansas woman who scolded me: "The only mistake I think your mother made in raising you was when she washed your dirty little face with soap she failed to wash your dirty little mouth." I promise to try to keep it scrubbed.

I'm also determined not to hold any grudges. Sixty priests, forty-seven nuns, and thirteen acolytes wrote to me: "You shouldn't carry grudges; you must learn how to forgive." And so I have been humbled. I will clean up my act and I will always love my neighbor.

Besides the people who are trying to change my speaking style, there's another pretty vocal group trying to change my smoking style. As all my friends know, I like a good cigar now and then. Smoking, of course, is becoming *verboten* in the U.S., at least in public, and lighting a cigar after dinner can almost touch off a riot.

Not long ago I was at Le Perigord restaurant in New York and I lit up a cigar after dinner. A woman whirled around in her chair and demanded I put it out. I was startled, but when I realized she was pointing at me with a lit cigarette in her hand I came apart. I also came close to butting my cigar on her nose.

On another occasion, I went to Nantucket with Kathi and Ned and walked into a bakery with a newly lit Havana cupped in my hand. I didn't take a single puff, but in less than ten seconds a kid in a white hat rushed out from the back of the bakery and said, "All right, who's smoking? Put it out or get out." I sheepishly raised my hand and went outside. I still marvel at that kid's nose. But what really humiliated me were those flinty New Englanders hissing at me as they came out of the bakery as if I were a pervert or a sicko.

I guess I'll have to go back in the closet as I did when I was a kid. This peer pressure is getting to me.

The biggest downside of fame is that I've lost my privacy—and I miss it.

Because of all the attention, I find myself becoming a little bit of a shrinking violet. I'm even resorting to strange disguises when I go out in public. When I was in Boca Raton recently on a rainy day, the kids and my mother thought it would be a lark to go to a doughnut shop that featured topless waitresses. I wore a baseball cap, dark sunglasses, and a beat-up windbreaker—but the disguise didn't work. A guy walked by, hit me on the back, and said he just loved his new LeBaron convertible.

Despite the little annoyances, I'd be less than honest if I didn't say that the net result of all this fuss is that people recognize me and like me. And I like that. You'd be a piece of stone if you weren't moved by all the young people who come up to say, "You changed my life. I'm trying harder in school." Or "I'm getting my act together because of you."

But the fame isn't going to my head. At least I hope not. Some guys grow and some guys blow as they get prominent and prosperous. It's the old story. You have to know that fame is fleeting, and I know—I always had my parents to refresh my memory.

No matter how important you think you are, they taught me, you're a mere nothing in the passage of time. Once you reach a certain level in a material way, what more can you do? You can't eat more than three meals a day; you'll kill yourself. You can't wear two suits one over the other. You might now have three cars in your garage—but six! Oh, you can indulge yourself, but only to a point.

One way to make sure fame doesn't change you is to keep in mind that you're allotted only so much time on this earth—and neither money nor celebrity will buy you a couple of extra days. Although I do have a rich friend in New York who says, "What do you mean I can't take it with me? I've already made out traveler's checks and sent them ahead."

Life is so complicated that it's hard for anyone, especially kids, to figure out what their purpose is in life, and to whom they're accountable. Of course, we should all be accountable to God throughout our lives—and live our lives that way every day, not just on our deathbeds begging for forgiveness.

A lot of people don't believe in God because they can't see him. I'm not a Doubting Thomas, though. I truly believe. When we were kids, our Sunday school teachers used to address this question by telling us: "You can't see electricity either, but it's there. Just stick your hand in the socket now and then to remind yourself." I've never seen an ozone layer or carbon monoxide or an AIDS virus, but they're out there somewhere.

It's always tough for me to define my own beliefs. I've never been a theologian in any sense, and so when I ask myself questions about this subject, the answers I come up with are pretty elementary. Do I believe in a life hereafter? Yes. How do I think I got here? I don't know, but I believe the Lord sent me. Why was I born in the U.S.? Lucky break.

As a kid, I was just an average practicing Catholic, and that's what I still am. I go to church on Sundays, but I miss every now and then when I'm traveling. Still, I've always understood the rites, and I've picked out a few that I find particularly valuable, especially the rite of confession.

I practice confession not because church rules say I should but because I feel good talking out loud about my sins and indiscretions.

I practice abstinence, or sacrifice, not because church rules say I should but because I like to test myself once in a while. I want to make sure that I'm still capable of denying myself, that I'm not hooked on anything.

For many years now, I have always given up cigars for Lent. Determined as I am, I must admit I've had some pretty close calls. When Chrysler was fighting for its life and we were doing deals late at night involving millions of dollars, I was nervous and just dying to light up. But I couldn't. I had made a vow to abstain for 45 days. (My kids tell me that if I were really serious, I'd smoke like mad during Lent and give it up the other 320 days.)

I have a friend, Bill Curran, who disciplines himself in much the same way, and he's not even Catholic. He loves Dewar's Scotch, and he drinks it in such huge quantities he once made a special trip to the distillery in Perth, Scotland, just to see what went into this magic elixir. But as addicted as he is to the stuff, every three months he goes on the wagon for two weeks. And he's never missed. He says it's the only way he has to test himself; if he ever breaks this vow, he'll declare himself an alcoholic and give it up, cold turkey.

Another of the rituals that mean a lot to me is prayer. Almost without fail, I pray every night. I started that routine as a kid, and I've kept it up all these years.

Why should I, a grown man, pray? Don't I have better things to do? Well, I think it's soothing to engage in prayer, to ask somebody out there who's more than your boss or part of your family for help.

I firmly believe what Abraham Lincoln once said: "I have been driven to my knees many times because there was no place else to go."

What do I pray for? I pray for good health for my kids, my

family, my friends, and myself. I pray that they don't meet with some terrible accident or find themselves in the wrong place at the wrong time. And I thank the Lord for all my good fortune.

I never pray to my God that Chrysler can make $20 a share this year. When the company was on the ropes, I did pray that we would make it through the day. But I'd never pray for a good quarter or anything as unimportant as that.

As I get older, I think more and more about what comes next. I know there's got to be something else after this life is over, because I can't grasp the alternative. I can't imagine that through all eternity I'll never see anyone I love again, that my whole awareness will just be obliterated. I can't believe that we're only bodies passing through.

When I muse about this, I think of all the great moments I had with my father. It's inconceivable that I had this wonderful period in life with him and then suddenly the curtain dropped. Instead, I want to believe I'm going to meet up with him again. I also want to have the opportunity to catch up with Mary, if only to tell her what I forgot to tell her, and to meet all my lifelong friends who have died. I do think they're out there someplace.

I haven't yet formed a clear idea about what the hereafter might be like. I don't know if everyone's an angel. Or an apparition. Or it's just all beyond comprehension. But I do hope that it's going to be better than here, because life on this planet is not exactly peaches and cream. I mean, this life is tough. I suppose that's the promise religion holds out. If you can take this life as it comes and give it your best, there'll be something better afterwards.

I know people who really believe in reincarnation. Others think that Shirley MacLaine is a kook. Out on the West Coast, people talk to crystals and listen to the sounds. Others go to séances and wait for the taps. Maybe we're all such innocent, frail people that we'll turn to anything that might give us some hope.

I've always marveled at how belief in the hereafter gets accentuated as people grow older. Until their deathbeds, many of the great minds in science thought that because their soul and being were wrapped up in their body—the old ninety-eight cents' worth of chemicals (before inflation, that is)—and that because after death there would no longer be a body, that was it. But now when they have to go, suddenly they want to believe in somebody up there because they don't know where they're going and they're scared—sort of scared to death, you might say. It's a little late by then.

That reminds me of the joke about W. C. Fields. He was a lifetime agnostic and yet he was discovered reading a Bible on his deathbed.

"What are you reading that for?" someone asked him.

"I'm looking for a loophole," he replied.

In the end, we do have some rules about what to believe in and how to behave. We've got something called the Ten Commandments. You'll notice they're not called the Ten Suggestions. They're not optional. They don't depend on your being in the mood. For example, I have a friend who says he's a good Catholic—but only when he's sick and when he flies.

If you don't know the Commandments, read them. It's humbling. They're pretty important reminders of human weakness.

They've served me well. But, then again, I doubt I could ever get too carried away. Any time my head starts to swell too much, I've always got my own kids to shrink it right back to size. Recently, Lia and one of her friends came by for dinner and Lia said, "You know, Dad, everyone says that you're a hero."

Then her friend chimed in: "Yeah, he's not only a hero, he's an Italian hero and that is one great sandwich! Let's eat."

PLAYING FOR BROKE

IV
GOOD BUSINESS—MORE ON MANAGEMENT

Whenever the subject of management comes up, everybody—and I mean everybody, right down to the janitor—seems to have some mystical approach to it. By now there have been about as many books written on management as there have been on diets—and, I might add, with about the same measure of success. There's probably even a grapefruit management book in the works somewhere.

How good are all those diet books? Well, none of them delivers unless you remember to do one thing: Don't eat so much. It's still the only way to lose weight. In the end, if you don't follow through on a few simple principles, what good is reading all those books?

The same is true of management. I got so much reaction to the management tips in my last book that I went back and reread them to see why all these testimonials were crossing my desk. All I'd done was toss out a few broad concepts—my quarterly review system, my firm belief in communications—and yet people were writing from all over the world to tell me they had turned their hardware store or their Good Humor route into a stunning success. I said to myself: I'd better try to figure out what my theory of management is. So here's my contribution to the never-ending debate.

* * *

If you make believe that ten guys in pin-striped suits are back in a kindergarten class playing with building blocks, you'll get a rough picture of what life in a corporation is like. Grown men in a meeting will do anything—absolutely anything—to avoid being shown up. If someone doesn't know the facts about a subject, he'll ad-lib, just like a kid. Instead of saying "I'll have to get that for you, boss; I don't have the answer right at hand," he'll try to fake it. He's scared that if he confesses he doesn't know, the boss will think he's not as sharp as the other little kids in class and maybe he'll miss nap time. As a result, he'll embarrass himself and babble like an idiot.

Only the boss can set a tone that lets people feel comfortable enough to say those magic words "I don't know." Followed by: "But I'll find out." Business, after all, is nothing more than a bunch of human relationships. It's one guy comparing notes with another: "Here's what I'm doing. What are you doing? Is there some way I can help you—and you can help me?"

Whenever I talk about this subject, I feel like a five-year-old myself. People are always saying to me: "But there's got to be something mysterious. There must be a formula." There really isn't. Start with good people, lay out the rules, communicate with your employees, motivate them, and reward them if they perform. If you do all those things effectively, you can't miss.

There are two broad management subjects that business people will argue about until long after I'm dead. One is the role of the staff (or planners) versus the line guys (or doers). The other subject is consensus management versus arbitrary one-man rule.

First things first. Staff, to reduce it to its simplest terms, is what supports the boss. I don't mean who brings the coffee, but who furnishes the information that helps the boss make decisions. The important question that every manager has to answer for himself or herself is: How much staff do I need to run my organization? In some businesses, often one good secretary will do. At Ford, either because of the family's mentality or the Harvard Business School mentality, you had to plan and analyze down to the last gnat's hair. Before you made a move, you had to examine every alternative and research every factor to be sure you didn't make a mistake. That was the ultimate sin. Ford is so chock-full of staffs that it even has a super staff—the Corporate Strategy and Analysis Staff—which oversees all the other staffs!

I don't care how successful Ford or anyone else might be with that approach; it's not the kind of environment that gives big business a good name. After a while, such companies have a hard time attracting young, entrepreneurial people, because there's just no room for guts management and instinct when you're loaded down with a lot of second-guessing.

My problem at Chrysler is exactly the opposite. I have a ridiculously small staff. My line guys are so aggressive that they may make multimillion-dollar mistakes before I've had one alternative to look at. Frankly, I'm so lean on staff that it sometimes scares me. That's why I've recently installed a few staff people, most notably Tom Denomme, who's listed on the chart as my vice-president of corporate strategy.

His real title should be Devil's Advocate. He tosses out an idea a minute, some of them a bit off the wall, in order to keep me in constant turmoil. It can get crazy sometimes, but I like it that way because I don't have—or want—a purchasing staff, a marketing staff, an engineering staff, and a manufacturing staff to ride herd on the operating guys who are doing all the work.

My feeling is that if I'm going to err, I'd always err on the side of leanness, because the decision making is faster. Of course, if you get too lean, you'll wind up making momentous decisions with no more information than what the weather is like outside. But you don't need a bloody bureaucracy to make sure everybody's ass (especially the boss's) is covered in case of a screw-up.

When it comes time to make decisions, you shouldn't get too old over them. Sure, they won't all be perfect. In fact, some of them will be duds. Learn from them, but don't stop trying. The introverts of the world, the non-risktakers, are probably that way because they got burned young. Maybe they made the wrong move in a marbles tournament or a game of checkers and now they're never going to take a risk again. That's no way to live—and it's certainly no way to make a profit.

The big fuss about consensus management is another issue that boils down to a lot of noise about not much. The consensus advocates are great admirers of the Japanese management style. Consensus is what Japan is famous for. Well, I know the Japanese fairly well: They still remember Douglas MacArthur with respect and they still bow down to the Emperor. In my dealings with them, they talk a lot about consensus, but there's always one guy behind the scenes who ends up making the tough decisions.

It doesn't make sense to me to think that Mr. Toyoda or Mr. Morita of Sony sits around in committee meetings and says, "We've got to get everybody in this organization, from the janitor up, to agree with this move." The Japanese do believe in their workers' involvement early on and in feedback from employees. And they probably listen better than we do. But you can bet that when the chips are down, the yen stops at the top guy's desk, while we're wasting time trying to emulate something I don't think really exists.

Business structures are microcosms of other structures. There were no corporations in the fifteenth century. But there were families. There were city governments, provinces, armies. There was the Church. All of them had, for lack of a better word, a pecking order.

Why? Because that's the only way you can steer clear of anarchy. Otherwise, you'll have somebody come in one morning and tell you: "Yesterday I got tired of painting red convertibles, so today I switched to all baby-blues on my own." You'll never get anything done right that way.

What's to admire about consensus management anyway? By its very nature, it's slow. It can never be daring. There can never be real accountability—or flexibility. About the only plus that I've been able to figure out is that consensus management means a consistency of direction and objectives. But so much consistency can become faceless, and that's a problem too. In any event, I don't think it can work in this country. The fun of business for entrepreneurs, big or small, lies in our free enterprise system, not in the greatest agreement by the greatest number.

Another thing that a lot of management experts advocate importing from the Far East is that the boss should be one of the boys. Democratic as that philosophy may sound, I don't think it's very practical. If the boss lets his hair down too much, he ends up like Rodney Dangerfield. No respect.

And yet, the boss can't be aloof either. A lot of the guys in the Fortune 500 seem to feel it's beneath their station even to talk to their own work force. Someone who's got 200,000 people working for him and who makes a million dollars a year begins to believe his position and his power make him infallible. He forgets to listen. He gets caught up in the clack of all the yes men around him.

My style, I hope, falls somewhere between those extremes. For example, I go to the National Automobile Dealers Association convention every year. In fact I haven't missed one in thirty years. Why? Because my presence tells the dealers in the best way possible that I think they're a vital part of the company. My being with them for a few days is the most effective investment I can make, and so I'm there religiously. I shake hands with the dealers and try to tell them how much I appreciate all their efforts.

By the way, that goes for everybody on the team. Once in a while you've got to show them you care. I have a minimum of four press conferences; I even care for the reporters (how about that!). I have formal management meetings with our top five hundred people four times a year. I visit our top bankers twice a year. You cannot call on these people only when you're in a jam. Handling crises is a hell of a lot easier if you've already got some rapport with the people who can help you solve them.

At Chrysler we've initiated an idea that works quite well. Every Monday, the top operating people have a meeting to go over basic operations. Before they discuss any business, they bring in someone from a lower level who has been named the winner of the week for his or her performance, and offer their congratulations for a job well done. Word gets around, and other people start thinking, *I'd like to be invited too.* They see that it's real recognition, right from the top.

Delegation versus the one-man band is another hot topic in management circles these days. The Harvard Business School says "Delegate," so people dutifully do it. But all too many of them never bother to get involved afterwards with the people they delegated to. This is where Reagan ran into trouble in the Iran crisis. The other extreme, of course, is the strong-willed leader who never lets go of anything, who has to be in on every decision. For instance, Donald Trump, the real estate tycoon, signs every check in his organization. Every check. He's a fanatic about knowing where each and every penny is going. Obviously, that style works for him, but in a big organization it sure can slow things down.

In the end, you've got to take a little of each approach. Alone with yourself, you have to look in the mirror and analyze your strengths and weaknesses. The stuff you're good at you can hang on to, but the stuff you're lousy at, you delegate. Then you try to learn from the person you delegated to.

As for me, I delegate plenty, but I still have a hard time keeping my hands off the marketing and design areas of the business. I want to be there, because I like them. And so I drive the people in those departments nuts. I've just got to stop carrying on this way!

Let's say you've done a good job delegating. Even if the people to whom you've assigned responsibilities are top-notch, you must let them know that you remember what you gave them and that you're keeping track, for everyone's sake.

Charlie Beacham at Ford was a great delegator. And by and large he picked terrific people. If I ever questioned some of his choices, he'd snap back, "Well, I picked you. What the hell are you complaining about?" But after he delegated, he used to drive his people crazy. He didn't need to know every last fact, but when he'd drop into your office, sit on your desk, and say, "How are you doing? I haven't seen you in three months, but your truck sales stink," you'd snap to attention pretty quick.

I used to label that "management by nagging." A lot of business people might say, "Boy, I wouldn't want someone like that over me."

But Charlie had such an engaging personality that he could get away with it. When he walked out, even though he'd just shoved a spear into you, you were still happy that he'd come to see you.

I learned that technique from him, and so I've always managed hands-on. If you ask my crew, they'll probably tell you that I'm a pretty good nagger in my own right. If I've got a fault, it's probably that I manage hands-on too much. You shouldn't get so antsy that your people don't even have time to find out where the bathroom is.

It's also very important to be flexible. I don't want to quote you the old cliché "Management's an art, not a science," but dammit if it isn't the truth. Some heads of companies maintain that they have a system and they don't care who they stick into it. They'll say, "Let's put No. 1573-8 in that slot," as if they're assigning a prisoner to a cell. I don't see how you can manage that way. You have to adapt to personalities or you're finished.

By the time people get to a company, they're pretty well molded. Time and again, I've tried to change people who are over twenty-one, and I don't think I've succeeded with a single one. Over the years, I've been stuck with people who had some rotten work habits, and so I thought I'd pump some energy into them. Although I'm a pretty good salesman and can often be very persuasive, they wouldn't budge. Not one inch. Why? Their parents and their grade school teachers got to them before I did.

Charlie Beacham was right on the mark when he told me: "Don't try to change anyone. Use your energies on something better. You might win over one in a hundred, but you'll take such a long time finding him that you'll go crazy trying."

What does that mean? You have to take people with all their warts. And then, to make the system work, you have to discipline them a little bit. You have to say, "Okay, I don't care how you grew up or what you are—here's the way we're going to run this ball club. And here are the plays. If you don't like them, it's going to show. By that time, you won't have to get off the team—I'll throw you off."

Once you've laid down the rules, you have to sit back and trust your people, even though you won't know for a while if they'll come through on the battlefield. You can never be sure with live ammunition if the lieutenant is going to take you up the hill or turn around and run like a scared rabbit.

That's why the worst threat to a company, I'm convinced, is when you take a chance on someone who winds up being in over his head. He doesn't know how to admit he can't cope, and as he screws up he ends up screwing up everybody around him.

On the other hand, you have to be able to gamble on unproven talent, or everything stagnates. One of the most important ways I keep in touch with bright lights is through the technique of "skip" meetings. I don't know who coined the term, but the idea is that you're skipping levels of management to chat with someone you normally might never hear from. This means that the chairman of the board gets to have a relaxed talk with someone several levels removed from him.

I began skip meetings many years ago, because the system had gotten so big that I was talking only to my top two or three people. Oh, I heard from plenty of people in endless committee meetings. But those meetings were so highly structured you could almost smell that the system had already filtered or homogenized what they could say. I didn't want to wreck the system or go around the organization, but I wanted to stop being insulated at the top of the pyramid.

So every few weeks I call in a department manager or a top engineer or a plant manager and meet with him one-on-one. These people are known as the high-po's—or high potentials—who, unless they mess up, will be running the company five or ten years from now. When they come in, they're usually a little reticent, praying they don't spill coffee on me or knock over a vase. I relax them by telling them this is not a performance review and everything's off the record and strictly confidential. Otherwise, they'd all run for cover. My questions are always simple. How do you get along with the rest of the system? Does it work? Do you know what's coming up and going down? Once I start firing away, I find that they loosen up pretty quickly.

If you handle things right, the idea takes hold. I'd been doing skip meetings for many years when suddenly I noticed some of my other top people doing them too. Now members of our executive committee have picked up the habit.

After a year's worth of skip meetings, I may not remember who told me what, but I manage to get a feel for whether things are going right and whether the engine's hitting on all cylinders. I also get to know, close up, a lot of our best middle managers, whom otherwise I would never even meet.

I'll give you an example. I had been trying to put in a system of brand management under which every brand of car would have one person accountable for giving that brand an identity to the public. The manager would have control of the Plymouth car line and its marketing, for example, a full three years before the product was due to come to market.

At first we only put our toe in the water. We appointed a brand manager for product and a brand manager for marketing, and we told them: You two are going to be like Siamese twins. Live with each other. Get together day and night.

Turned out, the idea was a fiasco. How did I find out? Because in a lot of the skip meetings, people told me it wasn't working. I heard the same things over and over again. There should be one guy, not two. Also, the current two weren't able to get access to the company's resources early enough to have real responsibility. They were stuck on the outside like cheerleaders. That's how I discovered that we had to take the system apart and give the brand managers genuine power.

Whether it's through skip meetings or other means, it's absolutely essential to let your people express themselves. And that means letting them make mistakes. You've got to allow them to walk into your office and say, "Boss, I blew it." That's called growing.

I'm reminded of the guy who says to the football player, "Geez, your team won the Super Bowl—how do you feel about it?"

"Terrific," the player says. "A dream come true."

Then the first guy says, "Yeah, but you didn't play. You were on the bench the whole sixty minutes."

"Well, that's true," the player says, "but I was suited up and felt like part of the team."

That's not my approach. My feeling is that there's nothing like playing. Being active is the key. I like to get my people into the game.

Last year I decided to reorganize Chrysler for that reason alone. I came to the conclusion that I wasn't using all my people to their fullest potential. I was deep in talent, but it was arranged in such a monolithic way that I couldn't get some of the second-level people into the flow of things.

That's when I said to myself: *I've got so much on my plate. Why don't I use some of these guys more? Why am I standing on the ceremony of an organizational chart?*

I decided to make a number of my second-tier people more accountable. At the same time, I also wanted to deploy them so that they'd be closer to the marketplace. I didn't need them to talk among themselves but to talk to the people who buy from us.

And so I divvied up the company into manageable pieces and told these executives to go play the game to the hilt. It's now their show. Make it or break it.

When you try something like that, you have to be cautious or you'll bruise some mighty big egos. In reorganizing things, I had to switch some areas of responsibility. One lesson I learned a long time ago is that once you've given a guy turf in a hierarchy, the minute you openly take away any of it, even if it's one lousy blade of crab grass, he gets miffed.

I suppose I can't blame anyone who reacts like that, because I was the same way in my career. When Henry Ford named me head of the Ford Division, I was given everything but the assembly plants. Now I didn't know a damn thing about assembly plants, but I did know that the guy I was replacing used to have them. So even though I was being given the opportunity of a lifetime, I was ticked off that Henry didn't have the confidence in me to give me the assembly operations. I needed those plants like a hole in the head, but I was offended anyway.

In my own company I've had situations where a guy's plate was so full that the food was spilling onto the floor, but if I lightened his load, he saw it as a threat to his power and forgot completely what the objectives of the company were.

I know right away when someone's mad, because he'll use the standard ploy to tip me off: You go to see the head of personnel and ask what your benefits are and what you would get if you retired tomorrow. Naturally word gets back to the boss. I have to confess it's a maneuver that I once used myself. The tricky part if you're the boss is that you can never be sure whether the guy's trying to deliver a message or whether he's just playing chicken with you.

In the last few years, a lot of people have come up to me and said, "Boy, you've got it made. You're powerful. You've turned a company around. You get paid like a rock star. You don't have to do a thing. You just come in and put your feet up on your desk and smoke your cigar." Well, if I ever looked at life that way, I'd be in a mess of trouble. Because you can never coast. You can never rest on your oars as the boss. If you do, the whole company starts sinking.

I know many heads of big institutions whose attitude is "I must protect what they handed me." Well, what you were handed has nothing to do with the present. Your legacy should be that you made it better than it was when you got it. Show me a chief who's happy with his lot and I'll show you a guy who's going to blow it.

I'll give you a perfect example. In March 1987, my top management met to vote on whether or not to buy American Motors. And the vote was against buying it. The majority of my executive group said, in effect: "Let's not do it. Life is too good now." In fact, they sounded a little like my mother!

When I heard that, it really got to me. I made a speech in which I said, "You guys are getting too old to be in this business. You're getting soft and maybe a little fat in the head too."

I heard everybody out and then I overruled them and told them that we were going ahead. I had faith in my ability and their ability to make AMC go. The point is, you always have to look for the next rung up the ladder, corny as that may sound. After a five-year comeback, Chrysler had settled down. We were beginning to get a lot of flab around the middle. In some respects we had operated better when we were poor and dying. That's human nature too. As CEO, I felt I needed something to get everyone excited again. If you don't keep challenging yourself, you start wasting away.

Not that you should shake the trees just for the hell of it—

simply to rattle the troops. I'm talking about the well-thought-out challenge.

My belief was that AMC fit our company perfectly. There were three big pluses: the worldwide fame of the Jeep name and product; a third dealer organization we could build on; and a brand-new state-of-the-art plant in Canada just waiting to be run by somebody.

Today we have a new objective to shoot for. We're going after 15 percent of the car market and 25 percent of the truck market. Those goals would have been unthinkable five years ago. Yet now we've got a good shot at accomplishing them in the not-too-distant future. Then we'll try for the next step.

I can already hear the skeptics out there saying: "Don't tell me, Iacocca, that you want to be as big as General Motors someday." On the contrary, we really believe in our company slogan. We don't want to be the biggest—just the best.

There's an area of management where I feel our company and many others have really blown it, and that's women in management. I'll be the first to admit that Chrysler has been as guilty as everyone else. Only recently have we started to get some superb female engineers. But now it's not easy for the opposite reason: demand. Institutions like MIT and Lehigh are training more women in processing and design, but those people are in such hot demand that there just aren't enough of them to go around. Every now and then we run across some women who are terrific, but when we try to hire them, my recruiters tell me that they won't come to Detroit. When I hear that, my answer is that those are just alibis and they're not trying hard enough. A lot of men don't want to come to Detroit, given their druthers, but if there's a challenge and the money's right, they come. And so will the women.

The auto business is still incredibly macho. Far too many men in town think that women wouldn't know the first thing about cars. Well, why wouldn't they? We've had a woman—Janet Guthrie—race in the Indianapolis 500, and she did pretty well. And Shirley Muldowney blew all the guys away on the drag strips.

The funny thing is that even though all of us in Detroit have told ourselves that autos are a man's world, we've got plenty of women on the assembly lines in the factories. So if it's a man's world, why are the plants so full of women? It seems that it's a man's world only in the thinking department.

Why do I mention this? Because part of management is picking the best people, and you shouldn't restrict your ability to pick the best. Yet that's what we're doing by excluding women. We talk about it, we debate it a lot, but the truth is that the industry record is really lousy when it comes to women in senior management positions.

Another big part of management is what I like to call the sting of accountability. In other words, to whom are you directly accountable. In a public company you're accountable to the people who own the company—the shareholders. They elect a board of directors every year as their representatives. As the chief executive, I'm accountable to that board.

One of the most important responsibilities a chief executive owes his board is to find an appropriate successor and to groom him, because you have to remember that at any time you could drop dead. I have a stable that includes Jerry Greenwald, Bob Lutz, Ben Bidwell, and Steve Miller. And I think that any one of them could replace me tomorrow and do as well as I'm doing. So I have a succession plan, and Chrysler has continuity.

What that means, too, is that a CEO has to know when it's time to step down. A strong leader in good health never wants to leave. He operates with blinders on about this issue, and when you mention retirement to him, he gets the shakes. He says that his products are in his blood and he can't step away. Harold Geneen built up ITT from nothing; his board practically had to take him out on a stretcher. Harry Gray is still on the board of United Technologies and probably won't leave except at gunpoint.

If your attitude becomes that you're in good health and the company can't do without you, then you've overstayed your welcome. Because you do grow older. You may be as capable as ever, but a time comes when you have to repot yourself.

I don't think that there's an exact moment when a leader should automatically let go, but there are signs to look for. Impatience is a big one. A short fuse is another. *Déjà vu* all the time is a third. Sometimes a young guy will come into my office all wound up over a new discovery. Little does he realize that I've heard that discovery fifty times. I sit there and think: *Why am I spending an hour listening to this nonsense? I could tell this guy the way it'll end up. Things don't change.* That kind of testiness is a flashing light.

An aging CEO is like an aging athlete. The good fighter knows when to climb out of the ring. He's slowed down a little. The aging baseball player knows that he can't run to first base anymore. Still, some ballplayers hang on by taking pay cuts of a million dollars. They say they have nothing else to do, that baseball's their life. Well, your heart may be okay, but hey, your legs gave way. To me, business is the same. Your mind doesn't go but your energy level does. You get a little stale.

I'm sure that there are people who think that I could never leave Chrysler. The truth is, I could leave tomorrow. I have plenty of other interests that excite me. When I go off to Italy for a vacation, I never think of Chrysler. But by buying AMC, I just fueled my batteries for a few more years, so I'm going to hang around and nag the guys a little longer.

And when I do go, I guess my only legacy as I turn off the lights and leave is to pin up on the wall my little commandments of management. Whenever my successor gets in a pickle, he might want to take a look at them. They're my distillation of forty-two years in the business world. And they're very simple.

1. **Hire the Best.** Nothing will make a CEO look better than a talented management team. When I'm asked about how I turned Chrysler around, I always make the point that I didn't do it by myself—a lot of smart, dedicated people did it. Actually, since according to *Time* magazine my ego is as big as all outdoors, I should probably take credit for having done all of it by myself.
2. **Get Your Priorities Straight and Keep a Hot List of What You're Trying to Do.** No matter how complex a business is, and ours is pretty complex, I believe you should be able to write down your

top priorities on a single sheet of 8½-by-11 paper. It's always amazed me to see how many companies, even small ones, devote hours of effort and literally tons of paper to detailed plans of what they want the company to do over time. There are a lot of different names for them—Long Range Strategic Plans, Ten-Year Business Plans, Five-Year Profit Plans, and so forth. I guess if you've got a big staff and lots of extra time on your hands, it's not going to hurt you. But I've never seen a long-range business plan that couldn't be boiled down to a single page of priorities.

3. **Say It in English and Keep It Short.** Everyone has seen examples of bureaucratic double talk in written communication. You know what I mean—a long-winded document that takes the reader through two dozen options and alternatives and ends up with any one of six or seven different conclusions. Most of us associate this phenomenon with government bureaucracies. But take my word for it, a lot of double talk exists in corporations as well.

There are three factors behind the mumbo jumbo. First, the almost uncontrollable desire to tell all you know on any given subject. Second, the love of adjectives and adverbs over nouns and verbs. And third, the desire to impress your audience with your depth of vocabulary. I once read a fifteen-page paper that was tough to understand. I called in the author and asked him to explain what was in the tome he had written. He did it in two minutes flat. He identified what we were doing wrong, what we could do to fix it, and what he recommended. When he finished I asked him why he didn't write that in the paper the way he'd just said it to me. He didn't have an answer. All he said was: "I was taught that way." And he was an M.B.A. to boot.

Write the way you talk. If you don't talk that way, don't write that way.

4. **Never Forget the Line Makes the Money.** The political maneuvering between staff and line organizations is a wasteful and costly exercise. Every chief executive has got to come to grips with how he parcels out authority and accountability between these two groups. So I have a single axiom that helps me remember how to manage these often conflicting organizations. When the chips are

down, it's the line organization that makes the money; the staff doesn't make a dime. I view the role of the staff as primarily to help the CEO do his job and to act as a catalyst to the line. If you really want to get a line group motivated, just float up a "staff idea" with the right amount of "Why didn't you guys think of that?"

5. **Lay Out the Size of the Playing Field.** I'm a strong believer in letting line operations "operate"—delegating to good people and then letting them do the job. But, you might ask, if the key managers are running the business, what's left for the CEO to do?

    I think a big part of my job is what I call "defining the envelope," or setting the limits within which line management can operate on a relatively freewheeling basis. It's similar to a parent telling a child: "Play in the backyard but don't go past the gate and don't climb over the fence and don't invite anyone over." The child has the run of the yard, but the parent has prescribed limits on where he can go and what he can do.

    As a CEO, I set limits. I set limits on the total amount of capital that can be spent—but not necessarily on how to spend it. I set limits on the number of executives I want on the payroll—but not who they are. I set limits on the amount of R & D spending I'm willing to support—but not the projects that are funded. And I establish the company priorities, which represent the limits or parameters that set the direction of the whole line effort.

6. **Keep Some Mavericks Around.** All CEOs should worry about getting fed a single point of view: that is, a party line that has been filtered, refiltered, homogenized, pasteurized, and synthesized—what we call "cooking the pudding." Without differences in viewpoint, and openness to constructive expression of these differences, a corporation can be led into a lot of bad decisions. There is a real risk in telling the CEO only what he wants to hear and never having any disagreements in his presence.

    To guard against this risk, I've always tried to keep some smart people around me who are contrarians, who for whatever reason will not accept very much at face value and are not

impressed by the rationale that something is being done in a certain way because it's always been done that way.
7. **Stay in Business During Alterations.** A lot of CEO surveys indicate that long-range planning and strategic planning are seen as among the most important responsibilities for a CEO. I'm not going to argue with that, but I've always felt that making sure you're maximizing earnings today is also a key responsibility. It's easy, I think, for an organization to get mesmerized by long-range plans. On paper, at least, they're neat and squared-off—and always work.
8. **Remember the Fundamentals.** I have always been a fan of Vince Lombardi, the late great leader of the Green Bay Packers. Although his teams had a fairly versatile offensive game, at least for that period of time, their real strength lay in their adherence to the fundamentals of playing solid football: blocking and tackling, good play execution, and mental discipline. They weren't the fanciest football team of their day, but to watch them run a power sweep with the linemen pulling out and blocking was to see a thing of beauty. It was also a devastatingly effective offensive weapon.

When all is said and done, management is a code of values and judgments. And that's why, in the end, you have to be yourself.

Which brings me to the best rule of management: Pick a style that you're comfortable with and stick with it. You can have role models, but don't try to be somebody else. Be yourself, stay natural and dammit, smile once in a while!

V
BAD BUSINESS — WHAT'S WRONG WITH WALL STREET

One morning as I was reading *The Wall Street Journal*, I came across some startling news. The paper was running two big stories about U.S. Steel and Goodyear being pounced on by a pair of corporate raiders. I said to myself: *Now wait a minute. Those companies are my two biggest suppliers. Without them, I don't make cars.* Yet here were these interlopers coming in and I was nothing but an innocent bystander. That was the moment when this merger mania really hit home to me. Things had gone entirely too far in Wall Street's version of Monopoly®.

Like all my counterparts in the business world, I was aware of the existence of raiders. And I knew they were adding a worrisome new dimension to the job of management. But I hadn't paid much attention to their antics because they were always doing their dirty work out of my sight. Now, for the first time, they had come to my meat house. Mr. Carl Icahn had dropped in. Sir James Goldsmith had stopped by.

Once they hit me where it mattered, I felt that it was time to offer a little perspective on who these guys were. Were they really Robin Hood and his Merry Men, as they claimed? Or were they Genghis Khan and the Mongol hordes?

All I know is what I see—and what I don't.

I see billions of dollars tied up in new corporate debt to keep the raiders at bay while research and development goes begging. I see billions going for greenmail, dollars that ought to be building new high-tech factories. I see confidence in Wall Street's integrity lower than at any time since the crash of '29. I also see a huge share of America's best management talent wasted on takeover games when it should be devoted to strengthening the industrial base of the country.

But I don't see the raiders creating jobs. I don't see them boosting productivity. And worst of all, I don't see them doing a single thing to help America compete in the world.

Over the last four or five years, I've watched the merger mania with a jaundiced eye. In the old days there was nothing wrong with mergers. There was a firm rule that you couldn't merge on a horizontal basis. GM, for instance, couldn't buy Ford. That would lessen competition. It couldn't even buy a big supplier, because that, too, would lessen competition. Then the antitrust laws had some teeth. But when the Reagan administration came along with its laissez-faire world, a new mood of "anything goes" took hold. When guys like Ivan Boesky heard the news, they started stuffing suitcases full of unmarked bills and going to town.

Today anyone can merge with anyone else. Antitrust is dead and buried. The result is a practice of merger for merger's sake, which means that since you don't have to worry about the consequences, you might as well put anyone who appeals to you into "play." That's the new jargon down on Wall Street. Another term for it is unadulterated greed.

Propelled by a lust for dollars, these opportunists decide that a company's stock is undervalued and round up capital—generally in the form of junk bonds, high-yielding bonds with hardly any worthwhile collateral behind them—and then stage a raid. But the whole thing is built on sand.

These new raiders are nothing more than people who have figured out that if you can rustle together enough money to take a position in a company, you'll end up holding all the aces. Because then you become an owner, even if it's in name only. The day you buy that stock you have the voting rights to threaten the current management that unless they play ball with you, they're going to lose their company—and maybe even their jobs.

Let's take a specific example. Sir James Goldsmith trotted over here from England and tried to buy Goodyear. He came well armed, with a $2 billion consortium of loan money, which is more than enough to scare the wits out of most managements. (Merrill Lynch provided the capital, with a preannounced $200 million fee for their services. See, everybody gets his share of the slops at the trough.) Goldsmith assured everybody that he was doing a wonderful deed by raiding Goodyear—he made it sound as if he were the Red Cross or something—because, he claimed, Goodyear was such a badly mismanaged company.

That's always the line these raiders give. They play on this myth that guys can rise to the head of big publicly held companies by being fat, happy, and stupid; that they're busy studying their golf scores and ignoring the poor stockholders. It's funny, but I don't know many people like that. If they're out there, they must be hiding behind their fern plants.

With Goodyear in particular, that scenario couldn't be further from the truth. In the tire business, it's the best-managed company in the world. I've known Robert Mercer, its CEO, and Chuck Pilliod, the CEO who preceded him, for a long time, and they're top-notch. And they happen to build the world's best tire. That's why Chrysler buys 85 percent of its tires from them.

As I watched the raid on Goodyear evolve, I started to get a little jittery. I knew that if Sir James got hold of the company, he'd start selling off assets as if he were holding a garage sale. Tires weren't what had piqued his interest; money was.

A popular conception, buttressed by the media, is that these raiders are so shrewd and have done such diligent research on the companies they're going for that they end up knowing more about their targets than anybody on the planet does. Well, here's what Bob Mercer told me about Sir James's sleuth work:

"The day after we found out Goldsmith was after us, I called him and suggested we talk. He invited me to a meeting at his East

80th Street town house in Manhattan. It included lunch, the most expensive lunch I've ever had. I hadn't taken too many bites before it became quite apparent that the stage was set to take over our entire company for its breakup value.

"It was obvious, Goldsmith said, that our assets were underperforming and undervalued. He would correct that by refocusing the business. In five minutes of discussion, I was able to determine how little he knew. For instance, he thought that our Aerospace division was a recent diversification. But we'd been in that business since 1911. He didn't understand what we were doing in chemicals. Yet the chemical business supplies synthetic rubber and other chemicals so that we can manufacture rubber products across a vast spectrum of product lines. He then questioned our involvement in the energy business, not realizing that there are seven gallons of oil in each passenger tire we produce."

Can you imagine Goodyear being run by somebody that ignorant of what he was buying? Well, Goodyear got lucky, because just as Sir James was preparing for the final pounce, the Ivan Boesky shenanigans were exposed and the stock market took a dive. At the same time, the Ohio legislature hurriedly put the finishing touches on some anti-hostile-takeover legislation. Goldsmith got nervous enough that Mercer was able to buy back his holdings.

Of course Goldsmith still managed to jet home with a cool $93 million in profit from his greenmail. Greenmail, in case you're wondering, is when a company pays a raider a premium for his holdings—if he'll go away. What I think it really is is blackmail in a pin-striped suit. Don't kid yourself. That's what it is—no more, no less.

As a result of that greenmail, Goodyear today is a smaller company. It had to hock everything it could to stay afloat. And I have to wonder how wounded my best supplier is.

What are the results of Goldsmith's aborted raid? Has Goodyear learned how to make a better tire? Is its research and development budget intact now that it took on this huge debt? What is the redeeming value of that attack? No raider has a good answer to those kinds of questions. All you hear is: "Well, we're not breaking any laws."

I'd like to ask any kid who's enrolled in Economics 101: "How do these hostile raids improve the productive capacity of the country?" That poor kid's got to be stumped.

After the greenmail was paid, Mercer told me, "We defeated Goldsmith's intent, but we were losers all the same. We had to sell off 12 percent of our revenue producers and take on a debt of $2.6 billion. Ironically, the one-year service on that debt is enough to buy a brand-new state-of-the-art radial tire facility in Cumberland, Maryland. Instead, we're closing down a plant in Cumberland and laying off 1,111 people. The money that would have purchased that plant is being fed to seventeen banks around the world to cover the interest on our debt. Add to that the more than $50 million in expenses it cost us in a two-month period for advice from investment bankers and lawyers, and you get an idea of the magnitude of the money wasted as a result of that raid.

"As a matter of curiosity, this particular raider earned over twelve times more from Goodyear in two months than I have in forty years of work with the same company. And he was never even on our payroll!"

The head of Safeway Stores, Peter Magowan, sits on my board, and he went through an equally wrenching ordeal. Safeway was a good company that was rolling along, growing every year, increasing its dividend, enjoying record profits, and yet it got knocked off. The company got hit with a hostile bid by this little drug company from Baltimore and was forced to do a leveraged buyout, a maneuver in which Safeway's management and some outside investors borrowed a pile of money to buy the company from its public stockholders.

It was like being hit by a bomb. To accomplish the buyout, Safeway had to sell off all its European operations. In fact, it sold seven hundred of its stores, cut its capital spending in half, and laid off a couple of thousand employees. That could wreck the company. Maybe all Safeway had going for it was good service. Or maybe it flourished because it had what people wanted: fresh food. Who knows what these reductions will mean? What if the company will have to cut down overnight delivery of lettuce? I don't know what you need to do to succeed in the food business, but I'm sure the guys on Wall Street don't know either.

You wonder how this can happen in America. It certainly isn't free enterprise. It's more like somebody coming out of the night and stealing your kids.

When people in other countries hear about all this nonsense, they must think we're a little crazy. I never read about a merger in Japan. I never hear of any leveraged buyouts there either. The Japanese probably don't even know what the phrase means. Over there, businessmen use wealth to create more productivity and to put more people to work. Here, nobody's building a better mousetrap anymore. They're trying to figure out how to do a leveraged buyout of the mousetrap company.

Along the way to making their killing, these raiders don't care if a company is decimated and people are thrown out of their jobs. Imagine Carl Icahn, up to his ears in TWA, wanting to try to get hold of U.S. Steel. He says the company isn't going to make it anyway, because it isn't modern enough and it's tried to diversify into gas and oil, so how much damage can he do?

Well, I know that type. He would go in there and sell off the steel business, because it's a capital-intensive business, while hanging on to the oil and gas reserves; that's all he's interested in. So the famous U.S. Steel (now USX) would be gone. And Bethlehem Steel would not be far behind. Then LTV, already in bankruptcy, would follow them right down the tubes. It's a helluva way to restructure the country's once-mighty steel industry.

Why should we allow these rugged individualists to rape the whole world? If you look in the history books, you'll see that we had these types before: the Robber Barons and the so-called railroad barons of the nineteenth century, who got this land for nothing and didn't care about anything but making money. But at least they left us some railroads and some prosperous industrial companies. Today's raiders simply shuffle paper. In a small world, in which we've got to compete with potent nations like Germany and Japan, we're going nowhere fast. If we were getting more productive in the process, then I'd say fine. That's what these raiders promise will happen.

"You wait and see, this will trickle down," they say. "We'll be leaner and meaner and better for it."

Is that so? The battle cry now is that the raiders are keeping yuks like me honest. Tell me how. I sure don't know.

I do know that if I spend $100 million on robots to improve my productivity, and I keep my costs down to compete with the Japanese, and I pass on $100 in savings to a car customer, then I've done something constructive. But if I take $3 billion in capital and go out and buy Burger King, you've got the right to ask me what in hell I know about hamburgers and french fries.

In observing these raiders, I've noticed a curious phenomenon. When they're on the attack, they shout their holier-than-thou pitch about making companies more efficient, liquefying capital, and defending the helpless stockholder. But a funny thing happens to these dedicated missionaries once you cross their palms with a little dough—they go away.

I don't mind when a company with a bad performance record gets taken over. Somebody ought to take it out of its misery—or carve it up. But that's not how these raiders operate. Maybe you've contributed to the defense effort, or you're providing the greatest hot dogs in the world, or you're rated one of the top five corporations every year; it doesn't matter. If there's a play, there's a play. And nobody cares what the raid does to the city, the state, or the country. If the raiders have to obliterate the company to get their money, well, that's business as usual.

If these raiders were really interested in saving American business from its incompetent management, where were they back in 1980 when Chrysler was flat on its back? One thing I never had to worry about in those days was a raider coming around, because Chrysler wasn't worth looting.

You see, the typical takeover target isn't a company in trouble. It's a company with a solid asset base, low debt, consistent profits, and a few bucks in the bank to help it get through the next business downturn.

When I went to school we called that "good management." Today it's called fair game. Choosing to modernize your factory instead of increasing your dividend might make shrewd business sense, but it's also like putting fresh blood in the water: a sure-fire way to draw the sharks.

No doubt about it, the raiders have pushed up stock values and made some people (mostly themselves) a lot of money. But if quick paper profits replace long-term competitiveness as the prime reason to invest in American industry, then I don't want to think about where we're going as a nation.

The bottom line of this mania is that if we don't watch it, we'll be heading for financial panic, because we're playing with dynamite here. In 1986 we had more bank failures than in any year since 1932. Even though I was just a kid in the early 1930s, I still remember all those banks going belly-up. My father lost most of his money in the Ridge Avenue bank in Allentown. I'll never forget it. He took me down there with him, and there was a mob scene in front of the bank. I heard my father say to one of the bankers: "What do you mean, 'bank holiday'? I want my money." The banker just looked at him and said, "Too bad. You can't have your money." It was frightening. I doubt if anybody wants to see those mob scenes again. You can bet I don't.

I'm particularly troubled by the deeper issue that has brought on this recklessness. And that's the deteriorating values of businessmen today, especially the freshly minted ones. It's time that somebody started to question the system that breeds these kinds of people. Why is it that talented young men and women aren't interested in going into industry but flock to Wall Street? Their only desire is to join some big firm, become a partner, and start doing deals. They're all hoping that by the time they're thirty they'll be making $5 million or $6 million and have their own airplane and even their own airport.

A lot of our best and brightest young people are being corrupted by the lure of Wall Street fortunes. The employment at investment banking firms and arbitrage firms has grown by leaps and bounds in the last few years. Try to get any of those people to come into industry. They laugh in your face. Although October 19, 1987, or Black Monday, may cut down on the laughs.

A worrisome attitude has taken hold among young businessmen.

It's called "I want a lot and I want it now." This young breed thinks people like me had our sights set way too low. I wanted to make a million dollars in my lifetime, and I thought that was dreaming big dreams. They want $50 million. And they want it this afternoon.

The conversation I hear in business circles is always about the deal and the play and the chase and the odds. It sounds a lot more exciting than talk about how we're going to compete or change the union work rules or return to common sense. Let's face it, those things aren't fun to read about over coffee in *The Wall Street Journal*.

These kids not only want to make bigger money, they also want to make it the easy way. Their attitude is: "I'm not my brother's keeper. Who said I was born on this earth just to create jobs for other people? I expect somebody to provide me with a job." Well, if everybody had that attitude in the past, there wouldn't be any jobs at all right now. There wouldn't be any companies to put into play either.

The anxiety of these young people to make money has risen to such a fever pitch that it was inevitable corruption would set in. Not only do you have dirty pool being shot, in the form of greenmail and the like, but you've got guys like Ivan Boesky and Dennis Levine surreptitiously moving the cue ball around. I've never seen so many business scandals in the papers as I have in the last few years. It's ridiculous. The financial pages read like the police blotter.

I just knew that once the possible profits reached into the billions this activity was going to get out of hand. I'm reminded of that old advice: "Don't spend time around a gambling casino. It brings in the worst kind of hoods in the world, because there's too much loose cash lying around." Wall Street has been converted into one big parimutuel machine. And it's brought in the seedy element.

I'm going to sound old-fashioned here, but to my mind the core of the problem is our permissive society. That's what makes kids going to college think that there's a difference between blue-collar crime and white-collar crime. They think white-collar crime is only a little bit illegal. Where I come from, a crime is a crime.

I never saw these values in people of my generation. I can't really give you a precise number, but I'm going to pick 98 percent. That's how many business leaders are guys like me who didn't inherit their wealth but worked hard, climbed up the rungs of the ladder, and after thirty years learned how to build a better garden hose or, in my case, how to build a better Mustang or a minivan. They put their nose to the grindstone and now they're reaping the benefits of that diligence. With rare exceptions, I find my peers to be good family men who are honest and ethical. Sure, there are some business guys who are dumb, but they usually get found out.

These young hotshots have forgotten how the system is supposed to work. It used to be that the purpose of the financial world was to serve the needs of business. In fact, that was its only reason for being. The financial system was to generate capital so that people could put it to work and create jobs and raise our standard of living. Now everything's reversed. The corporations have become the playthings of the financial market. The system has flip-flopped.

In early 1987, I called a meeting at Chrysler in New York at which I heard the unthinkable. Attending the meeting were our board, a few of our top advisers, and some of our biggest shareholders— one guy owned 10 million shares and another guy represented our biggest pension group, with some 7 or 8 million shares. The point of the meeting was to talk about defensive measures to avoid takeovers.

There was a lot of spirited discussion. I just listened and didn't say much, which was almost a first for me. After a while, one of the money managers made the point that his fiduciary responsibility to his investors was to maximize profits in the shortest possible time. If money managers didn't do that, he said, then they weren't satisfying the reason that investors gave them the money in the first place.

When I heard those words "maximize profits in the shortest possible time," I interrupted him and said, "Let me ask you a question. Let's take Goodyear Tire and Rubber [which was still fresh in my mind]. It's the premier tire builder in the world. The company's doing really well, but it's dependent on the U.S. car business, since the Japanese buy Bridgestone tires because it's a Japanese company. Suppose one day a big investor in Goodyear decided that he wasn't getting a fair return for his investment. Therefore he was going to raid Goodyear, throw out the old management, close some plants, put in $2.6 billion of debt, and own the company. Now he doesn't know a blessed thing about tires, and Goodyear might go to the dogs, because you can't take on $2.6 billion in debt and claim you've made the company stronger. But let's say short-term you could have made an extra five bucks by selling the stock."

I went through this litany and ended by saying, "So for a five-dollar bill you'd be taking a wonderful old company and destroying it. If anything, it would become less competitive, and then the Japanese tire companies that I'm worrying about would take over most of the U.S. car business. What would you do then?"

Do you know what the guy said? Without a moment's hesitation, he replied, "Hell, that's easy. The next day I'd buy Bridgestone stock in Japan. I'm not a social engineer here. I go where the action is."

I smacked my forehead and said, "Dammit, that says it all. That's what's wrong with the country today." To me, it was the clearest and saddest commentary on the state of American business that I'd ever heard in my life.

I'm looking to build that better mousetrap and to create more jobs. He can't spend a second thinking about those things, because dollar signs are all that interest him.

I'd like to play a tape of that meeting for the U.S. Congress and then ask if they really understand this competitiveness thing they're all talking about.

I should point out that I'm not opposed to all mergers. How could I be, when I did one of the biggest of all with AMC? But I go at them in a completely different way. At Chrysler, we say that we'll try to make the company efficient over time, but we don't plan to sell off assets on the first day. That's credo number one.

Nor do I go in saying, "I'm going to buy them but I can't afford to buy them, so I'll take on debt that will double or triple my interest costs. Then to service that debt, I'll give something up." What you usually give up is the long-term stuff—the R & D—which is precisely what you need the most if you're going to keep growing. Before I think about buying anyone, I first check my bank account to see that I've got the money.

My other unwavering rule is that I won't do an unfriendly merger. That's a mind-set of mine. Maybe somebody who succeeds me will attempt a hostile merger, but I just can't. To me, it goes against human nature, not just *my* nature. It's an aberration to come in and tell people that you're going to take them over whether they like it or not. That's being the schoolyard bully. I like the idea even less when you're not familiar with what the acquired company does. If you're in the business of building kiddie cars or Kewpie dolls, how can you possibly take over a roach-control company and then announce that you're going to lay off a lot of its people on the premise that its guys aren't as good as yours or else it would have bought you? There's so much tension in a situation like that, you could cut it with a knife.

So we're not going to kidnap anybody into the family. When it comes to courting, we like to do it the old-fashioned way—to knock on the front door with candy and flowers.

These days, though, with all the raiders running loose, everybody is scared to answer the door. Even a simple phone call can send a company to battle stations. And if you say that you'd like to stop by and talk, the top brass fills the moat and pulls up the drawbridge.

I found myself staring gloomily across the moat on several recent occasions.

The first episode I've dubbed "The Man Who Never Called Back." The company we looked at was Martin Marietta, the outfit that was involved in the first of the celebrated mergers, when Bendix tried to play out a Harvard Business School case study and attacked Martin Marietta in an unfriendly raid. Marietta reacted with a PacMan defense and attempted to buy Bendix. Then Allied rode in as a white knight and bought Marietta temporarily. The whole messy situation almost wiped out Marietta. And the end result was that a good company was saddled with a billion dollars in unwanted debt.

Chrysler was looking to diversify and get into high technology; of all the companies we reviewed, Martin Marietta was always in the top three on our lists because they had the best scientists. I checked the company twenty ways till Sunday, and it came up great in every respect.

We didn't know the management and we weren't sure of the best way to make contact. When we checked around, we found out that a lawyer of ours and Thomas Pownall, then the chief executive, had a mutual friend who was also a lawyer. So, in early 1986, we contacted that lawyer and he agreed to approach Pownall to see if he'd be interested in meeting with me. To our delight, he said yes.

I subsequently received detailed instructions on our meeting, so detailed they reminded me of a James Bond movie. I was to ask for Pownall at a specific time at the Madison Hotel in Washington, where he was registered under a false name, and then to proceed alone on the elevator. I was half expecting a peephole to open on his hotel door with the standard password request.

Well, we had our meeting and despite all the James Bond stuff, it was a good session. To my surprise, Pownall told me that Ford had approached him but that his guys had decided they weren't interested. Apparently, Don Petersen, the head of Ford, had walked in with an investment banker strutting along at his side. "An investment banker," Pownall said, "was like a red flag in front of me, because I saw those guys try to eat up my company once. I knew then that if they offered me the moon, I wouldn't sell."

But we seemed to hit it off personally, and after talking about how our two companies could come together, we agreed to meet again to look at the possibility in more detail. It was all very positive.

A short time later we met for the second time, in similar circumstances. I had my people prepare some information on a merger concept that stressed the operating benefits rather than the financial deal. That could come later. After going through the material, I think it's fair for me to describe Pownall as enthusiastic about the idea. If he wasn't, he was a hell of an actor. At the end of the meeting, he asked for a copy of the material we had reviewed so he could discuss it with a few of his key people. He stated quite clearly that after reviewing the package, he would call me and let me know where his company stood on the possibility of a deal. I agreed to wait for his call.

And wait I did. A week went by. A month. Several months. I

didn't want to look as if I were desperate for the deal, so I kept on waiting.

Then one day he called me from an army base and said he couldn't talk right then but that he'd get back to me. I never heard from him again. Not a letter. Not a note. Not a call. He simply dropped off the face of the earth. It was spooky. To this day, we all kid about it among ourselves: "Hey, did you get a call from Pownall?" "No, not yet."

What happened? I don't know; I may never know. I suspect he was scared that he'd get trapped again the way he had with Bendix. Maybe he thought that if word leaked out that Iacocca was seeing him, Ford might decide to mount a raid. The upshot, as far as I can figure it, was that Pownall was just too shell-shocked from the last ordeal. I can certainly understand why.

It was a little disappointing. In Ian Fleming's novels, Bond always wins and ends up with the girl. In this one, I didn't win and I sure as hell didn't end up with the girl—or the company.

My next encounter was with what I call "The Man Who Lost the Power." We had analyzed the impact of a merger between Chrysler and a high-technology company that had a pretty good fit with our operations. In this instance, I knew the CEO well enough to invite him in to take a look at the merger idea.

He readily agreed to a meeting. At this particular time, the CEO was nearing retirement and he gave me a strong hint that he would love to do one more big deal before he got his gold watch. I had my people put together a complete presentation on the merger idea. The CEO liked the concept, and after discussing it with us, he said he would explore the idea with his board and get back to me. Sounds familiar, doesn't it?

Well, unlike the man who never called back, this CEO did get word back to me. But it was sad in a way, because the basic message was that he no longer had the power to do such a deal. His board was about to retire him, and his successor had already assumed the real power in the company. A few months later the announcements came out and it was pretty clear we'd been talking to a lame-duck CEO.

The lesson here is simple: When you're talking deal, make sure you're sitting at the table with someone who can act. Otherwise, you're wasting your time.

Then I came up against "The Man Who Couldn't Come to Lunch." Not long after the Marietta overture fell through, we had identified Grumman—a medium-size aerospace company that looked pretty good to us in terms of operating fit and future business—as a possible acquisition. However, we didn't know a lot about its company culture, nor did we know the management group.

After talking about if and how we should introduce ourselves to the company, we decided that a gentlemanly approach would be for me to invite their CEO to lunch. I did just that. I wrote a letter to John Bierwirth, the chairman of Grumman, indicating that we were interested in buying his company and assuring him that if we accomplished nothing else, at least we would enjoy a good meal together.

The letter I got back from him was a classic. It must have been written by a team of lawyers, and it as much as said "You may have written us, but we're not acknowledging that you wrote us and we don't want to get together for lunch. I don't care who you are, Iacocca, don't link your name with ours, because it will create a problem. So get lost."

A simple no would have been adequate.

The moral is that because of all the bad mergers, good mergers are becoming impossible. Everybody's taken on a bunker mentality. Companies are paranoid today, and you can't blame them. Chief executive officers are trying to plan ahead while looking over their shoulders. Some of them are spending more of their time fighting off the raiders who are trying to take over their companies than they are fighting off the Japanese and Germans who are trying to take over their markets.

So what's the answer? There are a lot of ways to tackle the problem, but there's no doubt in my mind that we've got to put in some rules. The existing regulations simply aren't adequate to deal with today's business environment. Several years ago I was talking to Edgar Bronfman of Seagram's, who said, "If a merger accomplishes nothing whatsoever, it's just a play. Shouldn't we create regulations that say you can't deduct all the interest payments on these high-yield junk bonds? That's something that would stop these raids at the pass."

Edgar was a little ahead of his time. Now Senator William Proxmire is advocating that we take away the new shareholders' voting rights, which would have the same effect. Another worthwhile possibility is to clamp some sort of limit on the amount of junk bonds that can be issued. That would cut off the raiders' favorite means of raising money.

One of the best ideas I've heard is to create a waiting period before stock can be voted. Doesn't it make sense to insist that anyone who claims to be an owner must stick with his investment over the long haul? We make people register to vote in advance because we don't want even our dogcatcher to be elected by somebody passing through town on a bus. I don't see why American business should be controlled by people who ride through in the middle of the night with a fistful of stock certificates that they intend to sell first thing in the morning.

Let's make a rule that an investor must keep his stock for a year before he gets his voting privilege. I can just hear Sir James's response to that: "Oh, you're out of your mind. If you do that, I can't get in and out. It's too risky for me to make a hundred million dollars. I won't buy." Well, what would we lose if someone like Goldsmith doesn't buy?

Quite simply, most raiders won't tie up their money that long. They're not risk takers at heart. They bet only on sure things.

These rules may sound punitive. But when things don't work, you have to take action. Because if the financial system goes in a country, it's all over.

Unfortunately, new rules, whatever they are, can't weed out bad values entirely. More than on any law, we have to rely on the character and ethics of the people. We've got to tell our kids to get away from this fast-buck mentality and to start thinking about other people besides themselves. It's the only way for everyone to prosper.

VI
BUSINESS SENSE — AND NONSENSE

Back in the early seventies, at the height of the Vietnam War and all the campus protests, a major university invited a battle-hardened admiral and a buttoned-down business tycoon to speak at its commencement.

I was the businessman, and the admiral and I should have known better, because the popularity of both categories on campus in those days ranked just a notch above that of mononucleosis.

I've played some tough audiences in my time, including Congress and the Washington press corps, but none was tougher than those graduates. They were so hard on me, what with all the booing and catcalls, that for a while I stopped accepting any commencement invitations. *Who needs this shit?* I said to myself.

But the world seems to have come right side up in the past fifteen years. Businessmen are welcome on campus again, sometimes even with open arms. I get about 150 invitations a year and all I have to duck these days is an occasional champagne cork. Now the graduates don't want to tear the establishment down; they want their piece of it. And they're asking people like me to tell them how to get it.

If business leaders are going to carry this sort of clout with our youth, we'd better make damn sure we're setting the right example.

There's been plenty of talk in recent years about how corporate America is fat, bloated, and inefficient. According to this line of thought, business hires twice as many people as it needs, it pays its top executives about forty times as much as they need, and then it showers them with better perks than the king of Siam.

In fact, if you listen to some of the bureaucrats down in Washington, the only thing we fat-cat businessmen care about is how many zeroes there are on our paychecks and how low our golf scores are.

Well, I know plenty of chief executives who don't even know what an eagle is. But I do agree that it's finally time for everyone in the business world to face up to the question: How fat can you get?

All the highly paid, multimillionaire guys like me have to ask ourselves what our responsibility is. Is it always to make the next million? Is it always to push the stock up another two bucks so that our stock options shoot up as well? Or is it to look out for those who live in the towns where our factories are? To take care of the people whose life savings are invested in our companies? To help Americans who truly care about our country?

The most important thing a company does for a community is provide jobs. But its role doesn't end when the jobs aren't there anymore.

When I came to Chrysler, I had to close thirty plants over a period of a couple of years. It was the most miserable part of my job, though it did make me something of an expert on closing plants. Believe me, as the years pass, it doesn't get any easier.

Let me tell you about the Kenosha, Wisconsin, assembly plant, which Chrysler acquired when we bought American Motors. It's eighty-six years old and was once used for making bicycles and mattresses. The work force is great, but unfortunately the plant is antiquated and very inefficient. No less than 101 buildings make up the total complex. We just couldn't afford to keep it open. Business has a responsibility not to go broke, and so, reluctantly, we decided we had to close it down.

Naturally, that was a devastating blow to the 5,500 workers who would be losing their jobs, and to the town as well. This is the real pain of the so-called deindustrialization of America. I guess I'm the heavy here, but I feel the pain too.

I got hundreds of letters from workers and their families, some of them angry but most of them just filled with desperation about their kids' education and their mortgage payments. We had to go way beyond any legal obligations. We couldn't just leave town and say, "Last one out, turn off the lights."

So, we set up a trust fund of $20 million to help with housing and education. Then I explained that for two years all medical benefits would remain in place for them and their families. Then I explained that one in three, or eighteen hundred, employees would be eligible for substantial pensions.

Most of it fell on deaf ears. What they wanted was their jobs.

I felt awful that I couldn't deliver for them. I knew we'd raised their hopes and expectations when we bought their plant, but I told them the "fix" was not with me but with the legislators down in Washington.

We are trying to help one small community, but the bigger problem is that the federal government doesn't seem too troubled by what this represents. Lost jobs are the end result of unfair trade practices, which is a chapter in itself. We desperately need a national plan to assist people who suddenly find themselves without work. We can't just say, "Oh, we'll train them to flip hamburgers."

This is the thirteenth Kenosha in two years for the auto industry. What I want to know is: When does it stop?

It's never enough for a corporation to look after only its own employees. If companies want to be part of flourishing communities with educated work forces, then they ought to pitch in and help do something about it. Far too many never lend a hand unless they see some dollars coming in.

Despite some deadbeats, I'm pleased to see that corporate volunteerism is steadily rising. Allstate Insurance lends executives to the American Cancer Society and the United Way. General Mills spends millions of dollars to help the elderly find alternatives to nursing homes. Du Pont has some of its employees help ex-convicts find jobs. Honeywell retirees prepare tax returns for housebound senior citizens.

A few years ago, Atlantic Richfield "adopted" a public school in Los Angeles and sent its employees to classrooms to work as tutors and teaching assistants. One of the results was that the grades of the students shot up. So many other companies in the area were impressed by the effort that now more than half of the city's public schools have been adopted by companies. Firms in other cities have begun to follow ARCO's lead.

At Chrysler we're pretty active in the Junior Achievement program. Many of our people spend hours in neighborhood schools teaching youngsters about business and business ethics. A number of our middle managers set up fictitious companies with high school seniors, who decide on a product—a revolutionary mop, a new shaving cream—map out how to market it, and then see if they can earn a profit. It's a good way to learn early on what business is all about.

Despite these efforts, all the reports of greenmail and golden parachutes have made modern businessmen look like a money-mad bunch of pirates. In particular, there's been a lot of noise in the press about the scandalous sums executives get paid. Well, when it comes to paychecks, you've always got to test reasonableness. And times change. What was a whopping salary ten years ago just about covers the mortgage and car payments today.

In early 1987, I tried to hire a guy from Northrop to head up our new Chrysler Aerospace and Technologies Company. And I had to do somersaults and backflips before my board to allow me to spend more than $2 million, up front, to buy out his contract. It was sort of like buying out the remaining years on the contract of an ace pitcher with the Detroit Tigers, and my board wasn't too ecstatic.

We all found out the hard way that it wasn't too much money at all—because we didn't even get the guy. When he marched in to tell Tom Jones, the chief executive of Northrop, that he was leaving to join Chrysler, Jones flipped out. This guy was the number four or number five man at Northrop, and because of my offer Jones boosted him to president of the whole company. At forty-six years of age! So, often even big money can't lure away the real superstars.

I can't talk about executive compensation without a few words about my own. I made the cover of *Business Week* for being the highest-paid executive in America. In 1986, as I'm sure the whole planet knows by now, I received $20.5 million—which covers one hell of a lot of greens fees. When reporters asked me about the amount, I told them: "What can I say? It's like Everett Dirksen used to remark: 'A million here and a million there—before you know it, it runs into real money.' "

It's true that $20 million sounds more like Michael Jackson's income for a weekend than the annual pay of a guy who can't carry a tune in the shower. But you have to remember that all of that haul wasn't my salary. The bulk of that monstrous sum was from stock appreciation. My cash income—before taxes, that is—was $1.7 million, including a bonus of $975,000. That's a decent paycheck, all right, but it's not out of line in my cyclical, feast-or-famine industry. All the rest of the money was from stock options I'd been granted over the prior nine years.

Chrysler, you see, gave me paper nine years ago, when the company's stock was trading at a measly $6, and many people were betting it was on its way to zero. Those were the days when I was earning a buck a year and we couldn't even get the corner deli to carry us for a pastrami sandwich.

The financial picture has picked up a bit since those desperate days when we were looking death right in the eye. After two stock splits, Chrysler's shares soared to around $47. Am I supposed to apologize for that run-up? Am I supposed to bad-mouth the stock so it goes back down?

By the way, everybody who stuck with me since 1980 and held on to their stock has made a nice pile of money too. By the end of 1986, our total dividends and stock appreciation had added up to an unbelievable 860 percent, the best performance of any company in the Fortune 500. Even I am a little awestruck when I hear that number. And so the portfolio manager, the baker, and the hubcap maker—everyone who clung to their Chrysler stock—made the same 860 percent that I did. Alas, what goes up must come down, and since the crash of October 19, 1987, all that reported income in stock has dropped in half. I hope when they put me on the cover of *Business Week* next time, they'll remember to make the proper adjustment.

I remember a few years ago when Doug Fraser, who was then head of the United Auto Workers and a member of our board, voted against my bonus and my salary. After he cast his vote, Doug said to me: "Lee, this is nothing personal. I like you, and I think you're doing a terrific job for the company, but I've been with the UAW all my life and the most I've ever made is $75,000 for running that big institution. I don't quite think you need ten times as much as me. After all, how much pasta can you eat?"

That said a lot to me. Fraser is one of the most enlightened leaders in the history of the labor movement, and even he had the mind-set of a true socialist. Our system says that if you strive for excellence and you achieve it, then you get rewarded for it. At the same time, if the business sours, then it's an entirely different ball game. If my stats fall off—if my RBIs and my home runs go to hell and I'm having trouble with the bat—you'd be surprised how fast I'd be off the team. If you don't believe in doing things this way, then you ought to get on the next plane to Switzerland. Because those are the rules we play by here.

Which doesn't mean that I think our system of rewards is necessarily fair, or that there aren't abuses. But I do think that if you helped rebuild a company and beat the odds, you don't need to apologize for winning the game.

So I'm not going to give the money back (though I am going to give much of it away to charity). If I were forced to do that, then it might be time to dismantle the free enterprise system as we know it. Because if little kids can't aspire to make money the way I did—by improving the performance of their company and contributing to the economic well-being of their country—then what should they aspire to?

* * *

The situation isn't as black and white when it comes to business perks. A lot of mythology has grown up around what is or isn't a perk. For instance, I have a driver who takes me to and from work. To most people—especially the subway straphangers—that may seem like an unnecessary perk. Well, my days are so full that I use the commute to read documents—and my mail. (I've also been known to do an occasional *New York Times* crossword puzzle.) Otherwise I'd never find the time. I'd actually rather drive myself, because I miss testing the competition's cars as well as all our own new cars. But I've made a trade-off.

Buick recently decided to eliminate dining rooms for its lower-level executives. Now those people must eat in the cafeteria with the rank and file. Well, I think that move is more symbolic than a true cost-cutting measure. There's always some room for symbolism, but when I heard about it I thought about our own modest dining room. We feed all our officers there, as long as they pick up the tab for everything they eat. At a lot of companies the policy is that the executives eat all they want for free. That's the way we did it at Ford in the good old days.

As a rule, I usually don't even bother going to the dining room. I have a sandwich at my desk. When I do drop by, everyone inevitably ends up talking business. That banter drives me nuts, because I'm trying to swallow my food and relax. So I reached the conclusion that an executive dining room is no more of a perk than having a parking spot in the garage. (Now *that's* a perk, especially in the middle of a snowstorm in Michigan.)

The real problem is not the existence of perks; it's the abuse of them. I'm talking about freebies like trips to hunting lodges, or the use of a yacht—frivolities that involve charging the company for a life style you could easily afford on your own. When you see guys traveling to golfing matches in company planes, you really have to wonder if the company is keeping records and billing everybody, as it ought to.

If a worker is asked to mow an executive's lawn and trim the hedges, that's an abuse that's in a league with embezzlement. For those types, justice is swift—you get canned.

In most ways, Chrysler, like other responsible companies, is a tight and lean outfit. We're never going to live high off the hog, because we're Depression babies. In fact, the combination of being born during the Depression and going through the Chrysler mess has made me a died-in-the-wool cheapskate.

Like any profession, the auto business has its bad apples. But I don't think corruption is rampant. It would be too costly and too dangerous—and a company would quickly become notorious. Look at General Dynamics: When people hear the name, do they think of a sleek F-16 fighter? No. They think of how the company billed the government for dogs traveling first class on airplanes. It's going to be a long time before those guys live down stuff like that.

Overall, I really do think there was a lot more of this nonsense ten years ago. Today we've all been forced to become more competitive; we simply can't afford fancy frills.

If stockholders are going to pay top businessmen seven-figure salaries, then they ought to get their money's worth. I think they usually do. But there is one area where the public is being shortchanged, and that's in speaking out. There are plenty of able managers around, but when it comes to speaking their minds, you'd swear they were deaf mutes.

In a sense, business is just like baseball. Good ballplayers draw people to the seats. Some have charisma. Some, like Reggie Jackson, even know how to strike out gracefully. I used to love to go see that guy play; even when he never got his bat on the ball, he played with a fury. That's what chief executives have to do too. They have to stand for something. I don't think there are enough leaders in our executive suites, and if there are, you wouldn't know it, because it's so goddam quiet.

The establishment keeps insisting that a businessman should be completely faceless. He should always wear a three-piece gray flannel suit, and his face should never be recognized, not even by his mother.

Well, Americans truly want some folk heroes. The free enterprise system needs these mavericks, these catalysts, these mind-benders.

And I fall out of bed with a lot of my peers because they want to have it both ways. They want to be known as good guys but they don't want to speak up and take any risks.

Most CEOs want good PR, but they want it purchased for them by some flack. If a chief executive delegates the company's image—how the public perceives it and why—to some PR guy, he must be nuts. Yet most of them do. They issue a command: "Just make sure everybody says nice things about us, and when they don't, I'll buzz you."

You can't be out there in the workaday world trying to make a buck and hiding in the closet all the time. Of course, you also don't need to be as crazy as I am and do your own TV commercials, but there is a happy medium. I feel strongly that if you're not accessible, then you can't complain later on. If you don't want to participate in the dialogue, whatever it is or at whatever level it's taking place, then you've got no right to mouth off that you've gotten a bad rap because your constituents out there don't understand you.

For example, I think the adversarial ads that Herb Schmertz did when he was at Mobil served a useful purpose. Among other things, they were useful to Herb, because I think more people knew Schmertz than knew who the head of Mobil was! But that's still better than nobody knowing what anyone thinks in the corporation. Sometimes Mobil went a bit far afield, sometimes the company was merely grinding its own axe, but what the hell, it was paying for the ads.

Not everybody can afford paid campaigns like Mobil's. But there are other ways to speak out. There are those institutions known as press conferences, which a lot of CEOs seem unfamiliar with.

I don't mean to needle my peers, but if the heads of the Fortune 500 companies were more outspoken, we'd all be better off. Far too many CEOs consider their companies private duchies of their own; they don't want to get involved. I tell my friends all the time: "Hey, why don't you speak up?" They don't answer. I don't think they know how to speak up. Maybe they need to hire a tutor.

Chrysler takes a position on practically all economic issues. Maybe we're nuts to do that. But I think that's wiser than the majority of companies, which come out of the closet only on the provincial issues that affect them. What's really lamentable is the fact that so many businesses don't give a damn about what's going on down in Washington. Their attitude toward government is "Who needs it?" They believe that you should stay home, mind your knitting, and let someone else worry about the problems of the world.

But your business can sometimes become everyone's business when you're trying to sell a product. In the last year, I made a couple of controversial decisions in the advertising world. Chrysler had shelled out $6 million for commercial time on *Amerika*, the fourteen-and-a-half-hour miniseries in which Russia manages to take over America. There was a storm of protest from various quarters before the show aired, complaining that it was anti-American and unfit to be shown. When the complaining got hot and heavy, I took a look at a tape of the first four hours of the show. The next day, I canned the whole show.

But I canned it not because I'm a rightist or a leftist but because the show was lousy. When I watched the tape, I actually thought Kris Kristofferson, one of the stars of the series, was falling asleep delivering his lines.

I've always been a big believer that if you're going into people's living rooms or bedrooms to entertain them, there should be something to keep the viewer glued to the set. I'm trying to sell a car during the breaks in the entertainment, and *Amerika* just wasn't a good way to sell cars. When I tuned out after a couple of hours, I figured everybody else would tune out too.

Of course I didn't say any of this at the time, because I didn't want to embarrass ABC before the show even aired. But I knew we had made a mistake, and when it's a big mistake, I like to cut my losses.

Not long after we pulled out of *Amerika*, the show *Escape from Sobibor* aired. This was a show about the escape of Jews from a German concentration camp. It was so gripping that I couldn't stop watching it. We were a heavy sponsor of that show, and I was glad we were.

Immediately afterward, however, I was inundated with letters from Ukrainians. They said they were outraged that Ukrainians were used as guards in the concentration camps and were shown shooting escaping Jews. In fact, they were so offended that they even threatened a boycott of Chrysler products—and, to top it all off, sued us.

This time, I was completely puzzled. I hadn't at all felt that the Ukrainians were shown as the heavies in the concentration camps. After I got a few thousand letters, I went back and looked at the film again. Sure enough, there were Ukrainian guards at the end shooting the Jews escaping from the camps. But the script was true to history. So I defended that purchase, and I would buy those ads again. Incidentally the show won a Golden Globe award for the best drama special of the 1986–87 season.

You simply can't bow to pressure if you think you've made the right decision—or you'll be too paralyzed to act. Any time parents complain to me that I've got a commercial on a show they feel isn't good for their kid, my reply is: "I'll take a look at it." If the sex and violence get out of hand, we pull out as sponsors. But it's tough for a businessman to play censor to the world. Often, the very kids the parents want to protect are at an X-rated movie that night. Parents should be the real censors. You can always yank the TV out of the kid's room, can't you?

All of this boils down to: If you take no risks, you do nothing. I can't add up the favorable and the unfavorable letters and then act according to a poll. If the goal is to satisfy everyone, then I'd better get out of the game. Don't be in business; just stay home under the bedsheets and read comic books. If I listened to every constituency and knuckled under to every pressure that came along, I'm convinced I would never do anything. That would ease my workload quite a bit, but what would I accomplish?

As sheepish as many companies are, they still manage to veer off in all sorts of ridiculous directions. I can't understand the recent trend of corporations' wasting a lot of their management's time and their shareholders' money in order to change their names.

United Airlines blew something like $24 million to convert all of its signs from UNITED to ALLEGIS. How did that happen? Well, some consultant waltzed in to United and announced, "In case you haven't noticed, you're a diversified company now, boys, so you need a new name. And we've got just the one for you."

Didn't the United management stop to think that it takes around fifty years to build up a good name like IBM or Coca-Cola or United? Unless somebody put a gun to their heads, why would they want to change it? Especially to something as wacky as Allegis?

But the name game is going on all over. The old Burroughs-Sperry is now Unisys. U.S. Steel has become USX. Anyone looking for American Can better check under Primerica. Some of the other new names are Omnicom and UNUM. Nobody has the faintest idea what the hell those letters stand for.

One of the fancy consultants who cook up these names even dispatched a couple of experts to take a close look at our name. After doing whatever it is they do, they suggested it was essential that we change it to something like Chrysco. Their rationale was that Chrysler is identified with cars only, and now that we were in the aerospace business and into financial services we rightly ought to have a new name. And Chrysco, they assured us, was a terrific choice.

I looked at them in wonderment and said, "Are you guys for real? We're a respected company now. Why would we want to throw away the name Chrysler for something as stupid as Chrysco?" People would think we had gotten into the shortening business. All I could say was, for "Chrysake" leave the name alone!

What really blew my mind was that some of my own guys were gung-ho to change the name too. When they told me that, I promptly kicked them out of my office. It just goes to show that if you have enough counselors advising you, you'll spend all your time missing what the real priorities are in life. I should be trying to build a better product, not monkeying around with silly things like new company names and logos. Next thing you know, some marketing genius will want to change my name to Coco, because it might play better on television.

I always have to scratch my head when I see some of the pronouncements coming out of corporations. Whatever you do in life—sell advertising, make bread, build cars, you name it—you've got to want to be the best. But there's a difference between being the best and just claiming you're the best. These days there seems to be a lot more talking than doing. Whenever I muse about business hype, the example that immediately pops into mind is GM's Saturn project. There was the hype of the century.

As it happens, I think Chrysler may have been partly responsible for the PR monster that Saturn became. For years, GM had been saying that it believed in free trade, but while espousing that philosophy it quietly made arrangements with three Japanese companies—not with one, as we did—to import up to a million cars into this country. When I first heard that, I yelled "hypocrite" and complained that GM had given up on the small-car market.

So we stung them a little, and in our own obnoxious way we kept raising the volume. It got the GM guys so mad that they hastily put together this Saturn program from scratch which would take on the Japanese with a world-class little car built entirely in the United States.

Well, the words were barely out of their mouths before they realized they had stumbled upon the PR coup of all time. Because the press danced a jig over it. The details were blared from coast to coast: Saturn will be a $5 billion program; Saturn will produce a revolutionary new car; Saturn will have the President of the United States ride in the first prototype.

As I listened to this avalanche of propaganda, I said to myself and to the press: "Now wait a minute. P. T. Barnum might go for this, but this car will never see the light of day."

"Sour grapes," they yelled back to me.

I agreed, and I was envious. Not because I thought the car was great, but because as a huckster myself I hadn't thought of the PR scam first.

Still, in short order Saturn took on a life of its own. GM needed to find a place to put the Saturn plant, so the company staged a bidding contest with every state in the union to see who would give it the biggest tax concessions in exchange for landing the new facility. It was unbelievable. GM had every state and all the governors drooling to try to give away the most money. Four governors even showed up on the *Donahue* show to beg the nation to support them in this runoff.

By the time the scramble was over, Tennessee was leading the generosity parade. So GM announced that it was going to build the plant in Spring Hill, Tennessee. Then it was going to build a foundry. Then it was going to build a stamping plant. There were going to be two production lines. There was going to be modular construction. I said, "Holy smoke, this goddam car is going to fly."

At about this point, I think the GM brass began to believe their own press clippings.

While all this was going on, I kept reminding the press that we had a program like Saturn called Liberty. We'd started it before GM did, but we hadn't wheeled out some prototype and gotten the President of the United States to ride in it. Our concept was that we would try certain inventions and innovations from that project and incorporate the ones that clicked into other car lines over the next five to ten years.

GM, however, made it sound as if its advanced engineering project would be an actual Saturn car. To show you how absurd things got, GM began to put out press releases saying the advertising for the car would be created not by the guys on Madison Avenue but by the workers. Now I thought everybody was going mad. No two guys in our company can ever agree on an ad. And they were going to use ten thousand.

Sure enough, little by little, the Saturn project is coming unglued. After the press reported on talk inside the company about dumping it completely, GM announced that Saturn would no longer be a small car. It may, in fact, be a very expensive large car. There will no longer be a foundry and a fully integrated plant in Tennessee. The anticipated volume has been slashed in half. And the Chevrolet dealers are expected to sell the thing.

In effect, GM is simply announcing another Chevy for 1990.

The free ink that was written on Saturn and the kudos that were handed out were enough to kill a grown man. And now it looks as if the great Saturn coup of all time is dwindling into a big nothing. It was just razzmatazz.

What does that say to us? What it says to me is that you should never do anything just to impress the public. Everybody should learn from Saturn never to get ahead of your interference when you're running the ball.

Saturn, unfortunately, is just one of GM's headaches—and they're all Excedrin ones. The company continues to believe that Alfred Sloan laid down the gospel, like God, and it can never change. But the world has changed. It really has become a global market, and GM's executives have never bought that concept.

One thing I know is that the GM natives are getting mighty restless. In the spring of last year, Victor Potamkin, one of the biggest GM dealers in the world, called and said he had to see me; it was very important.

Potamkin is an old friend of mine and a real megadealer. He thinks big—I mean, gigantic. He's built up an empire of about fifty dealerships, including a couple of Chrysler and Dodge franchises. When I met with him, he told me that the GM dealers were on the warpath and didn't think that the company had a game plan, that it didn't have the right products and wasn't able to take them into the twenty-first century. After talking to a lot of the larger dealers, he said the consensus was that if I took over General Motors, I could save them.

It was a good thing I was sitting down. I looked at Victor and told him he had to be out of his mind. "Are you kidding?" I said. "That's almost worse than taking over the United States government. I think big sometimes, but not stupid."

Well, then a funny thing happened. Ed Hennessy, the head of Allied-Signal, came to see me. We were chatting about doing something together in the aerospace field. In fact, we had both dropped out of the bidding for Hughes Aircraft at $3 billion. (GM eventually won the bid for $6.2 billion.) So while we were talking about Hughes, he said, "Why don't we buy GM together?"

Again, I darn near fell out of my chair. For one thing, I'm against unfriendly takeovers, and with Ross Perot gone from GM's board, you could be sure this would be a world-class unfriendly takeover. Can you imagine a canary swallowing a cat this size? But Hennessy started running figures by me, saying that if he could take over all the parts-supply companies and I could take the five auto divisions, the finance company, and the dealers, we'd need only $40 billion. It reminded me of the story where the guy said he bought the Empire State Building for $250 million, but the deal fell through. When asked why, he said, "They wanted a hundred dollars down."

I checked my wallet, and it was a little short of $40 billion. The whole thing still sounded wild, but with some shrewd borrowing, it might have been possible. We sincerely thought we could reorganize the system and use all of GM's best people to propel them ahead of their market rather than behind it. *Who knows*, I thought, *maybe the idea isn't so screwy?* After all, if somebody didn't come along from the outside, it might take GM years to restructure from the inside.

We took one more step. We arranged a secret meeting with Felix Rohatyn, who acts as an adviser to both of our companies. He was pretty excited too. He even suggested that we contact Ross Perot, who usually has a few bucks lying around, to see if he wanted to pitch in.

I went back to tell my guys about these meetings. "You know, fellows," I said, "we had a big argument here about buying AMC; it was a split house. But the ink is hardly dry on that and I'd like to move on another deal—almost as big. We're going to buy General Motors."

Everyone just coughed a little—and then their eyes glazed over.

For the time being, however, GM can relax. After meeting with Felix and Ed, I decided to check with a takeover lawyer on the sly to see about the legalities of the whole thing. Other issues arose, too, such as how do you come up with really big money? In the end, I concluded that it might be easier to buy Greece.

I sort of chuckled to myself, though, that in this day of no antitrust laws and anything goes on Wall Street, almost anyone could actually take a run at GM today and have to be taken seriously. And that's a tragedy.

Perhaps the biggest responsibility of any corporation is to own up when it makes a mistake. I know a little bit about this subject, because last year Chrysler goofed up but good.

It all began as just another day in the life of an automobile executive. I walked in on the morning of July 23, 1987, and before I could even have a cup of coffee, Jerry Greenwald greeted me with the news that we were going to be indicted the next day by a federal grand jury in St. Louis. And the Justice Department was going to hit us for $120 million in fines, and file criminal charges against two of our executives, Frank O'Reilly and Allen Scudder. I did a double take and said, "You gotta be kidding!"

The problem, he told me, was that we had been testing cars at a couple of our plants with the odometers disconnected. What's more, a few of the cars had been banged up during the tests, fixed, and then sold as new to customers.

As Jerry told me this, a bell went off in my head. In the previous October, as he and I had been touring our St. Louis plant, he'd mentioned that one of our guys had been stopped by a state cop for speeding. The guy had tried to weasel out of a ticket by telling the officer that he didn't know how fast he was going because his speedometer was disconnected. The police, believing that to be highly irregular, got suspicious and an inquiry was launched—unbeknownst to me.

At the time, the incident didn't strike me as any big deal. That shows how old I'm getting, because my antenna didn't go up. It's a common practice in the auto industry for employees to drive new cars to test for defects. As far as I knew from my Ford days, the testing was usually done with the odometers disconnected so that we wouldn't be short-changing the eventual buyers on their warranties.

I blame myself for not looking into the matter. Back then, if I had said the magic words "Are we regularly running cars with speedometers disconnected?" and been told yes, then my next question would have been: "What are the other guys doing?" And if they had said that both Ford and GM had stopped the practice years ago, which, by the way, we still don't know for sure, I'd have said, "Then why in the hell are we still doing it?"

But I didn't ask the key questions, because as I age, I must be getting a little soft in the head. In this instance, my experience backfired on me. At Ford, I must have test-driven a million miles in new cars with the odometer cable dangling under the dash and the radio antenna strapped to the windshield so we could listen to music on the way home. Those were the last two things we hooked up before delivering the car. It never dawned on me that the practice had changed.

That was the last I heard about odometers until I walked in and Jerry told me of the incident. I was mad—but only at myself. The boss is always supposed to be on top of his team, but he's not perfect. Unlike the Pope, he is fallible, believe it or not.

The other guys who really should have known better were our lawyers. When that employee was stopped by the police, our plant manager had called one of our lawyers and asked him if he could continue to disconnect odometers. He'd said, "Yeah, we've always done it that way." In retrospect, that was really lousy advice, even if we weren't breaking any laws. Lawyers, especially cautious by nature, usually get six other opinions on matters like this. This time, zilch.

The indictment got big play in the media—and the stories made us look pretty bad. The New York *Daily News* dug out what must have been the worst picture of me they could find and ran it under the headline WOULD YOU BUY A USED CAR FROM THIS MAN? YOU MAY HAVE.

Deadly, huh? That headline was picked up all over the country and set the tone that the guys at Chrysler were nothing but used-car jockeys.

Sitting around sulking, though, wasn't going to get us anywhere. We were being indicted, and we needed to take some action—pronto.

The next day, while my guys were still reviewing the case, I was in Washington for one of my frequent visits with Congressional leaders. That night I had dinner with Sam Nunn. Sam and I were supposed to talk about things like reflagging Kuwaiti tankers and his recent meeting with Gorbachev, but before we could get to those subjects he said we had to talk about important stuff. "Hey, Lee," he said, "how about this odometer thing?"

I thought: *Holy smoke, here's Nunn, the head of the Armed Services Committee with lots on his mind, and he wants to talk about disconnected odometers.*

That wasn't too bad. The following morning, I went to see the majority leader of the United States Senate, the Honorable Senator Robert Byrd, to talk about pending trade legislation. We discussed campaign financing, the trade bill, the legislative process—you name it. After an hour I said, "Well, I know I'm holding you up and I have to get to the airport."

And he said, "Oh no, I've got a couple of things to cover with you. Very important." He went over to his desk, got a cigar, and said, "Lee, I'm concerned about my Chrysler and its odometer."

My jaw dropped. I said, "You are?"

He went on: "I just bought it—a Fifth Avenue—in Florida, and my wife and I brought it up to our home in West Virginia. It's got eight hundred miles on it and I don't think the trip is quite that long. I'm a little worried now. So here's my serial number—could you check out the car to see if I have anything coming to me?"

I almost fainted.

We did check it out, and his car hadn't been in the testing program. But the incident hit me like a ton of bricks. I thought: *What the hell is going on? If the Senate Majority Leader is so concerned, then this thing must be unbelievable out there.* I flew back to Detroit and immediately called an Executive Committee meeting. "We've got to do something fast," I said. "The integrity of this company is on the line."

The first thing we did was review the facts. The worst part of the case was the notion that we had sold customers damaged cars. Over a ten-year period, there were seventy-two cars that had incurred some damage. Thirty-two had been banged up enough that they were scrapped or put into company service. Forty had been repaired and certified as okay.

The average damage was—get this—a hundred dollars' worth. In this day and age, that's a scratch. When we build cars on the assembly line, there is a lot of what we call "in-process damage." We take those cars to the repair bays and fix them. Those costs are huge in comparison. With the test cars, here was a typical example: One of our guys had made a left-hand turn into a Kentucky Fried Chicken place to get chicken for the kids and was struck in the taillight. That was the extent of it.

Unfortunately, of the forty cars that were later sold, there was one—one lousy car—that had gone into a gully and turned on its side while it was being test-driven. The press reported that it had flipped over, although it hadn't. But that car became the famous flying Plymouth Turismo. The press made it sound a little like the Flying Zamponis in the circus. The repair bill on the Turismo had been $950, since a quarter-panel had to be banged out and some of the trim replaced. In this instance, there was no doubt that the car should never have been sold.

Next, we looked at how much mileage had been put on the cars in the testing program. The rules were that if a guy lived twenty miles from the plant, he was to go home and come back the next day, putting on forty miles, tops. If he took a car home on a Friday night, he was to lock it up over the weekend and then drive it back Monday morning.

On paper our great crime didn't seem to be all that terrible. As far as we could determine, the most mileage anyone had driven was 418 miles. There's no question that was far too much. The next atrocity, though, was down to 185 miles. The next one was 165, and the next one was 85. So of the thousands of cars involved, there were only four that well exceeded our mileage limits and only one that had been badly damaged.

As far as the public was concerned, however, it wasn't the facts of the case that were uppermost; it was our credibility. And it had been hurt, badly.

Our quickie survey of customers told us they were at the high blood pressure level. The survey showed that just four days after the indictment, 69 percent of the population over age twenty-one knew about the case, and 55 percent of that group felt that we were bad boys. A company can only exist as long as the public is satisfied with its performance, and the public was telling us it didn't like what we'd done.

The damaged cars, I told my guys, were a stigma of the first order. So I said, "We have to buy them back, no questions asked." A customer might have driven his car into the ground; it didn't matter. He would get a new car of comparable value. As for the cars that had their odometers disconnected, I said we had to stand behind them. We'd tack on two extra years on their warranties. Then I wanted to throw in some overkill, so I said, "Let's give them two years on the power train warranty, which is worth several hundred dollars, and let's give them the full warranty that covers all the major systems of the car, which we sell for about $250." Some of my guys complained that we were overdoing it, but I told them: "Hey, at times like this, don't quibble."

Now came the hard part. After we put that program together, the advertising people came to my house on a Sunday and we wrote an ad announcing it. I was pretty pleased and hopeful until I came in on Monday morning and my lawyers were waiting for me. "We saw your plan," they said. "You can't go through with it."

"Why not?" I said.

And they replied, "Because we'd have to get permission from the U.S. Attorney in St. Louis, and we've been told that it would be construed as tampering with evidence if you bought back those forty damaged cars."

I said, "Hold it. I'm not destroying the evidence. I'll buy back the cars and they can impound them and use them for whatever they want."

"That makes sense," the lawyers said. "But it goes beyond that. There's a thing called obstruction of justice. And even if the government gets that evidence, the fact that you are going to a customer who could be a government witness against us would be compromising him. In effect, you'd be influencing his testimony."

I flipped when I heard that. All I wanted to do was take care of my customers. But the lawyers were saying I couldn't. I felt I had an obligation after making a mistake to tell the people I'd offended that I was sorry and to make amends. And they were telling me "No way."

That was a really bad day in my life. I went home half-sick. I couldn't eat or sleep. I came to the realization that our whole country was drowning in legal bullshit.

The next day, Dick Goodyear, then our chief legal counsel, came up with a solution. He suggested that we tell the U.S. Attorney what we were going to do and let him try to stop us. Goodyear's thinking was that the U.S. Attorney wouldn't be brave enough to say "I don't want you to take care of the people who were injured."

He turned out to be right. The U.S. Attorney told us, in holier-than-thou tones, "You have to decide for yourselves. However, remember that we in Justice are always on the side of making amends to anyone who has been victimized by a crime."

What crime?

So we proceeded. On the day that we ran our ads, I called a press conference in Detroit to offer a personal apology. I told everyone that test-driving cars with the odometers disconnected was plain "dumb." Selling cars that had been damaged, I added, "went beyond dumb all the way to stupid." I said it was unforgivable and it wouldn't happen again.

It was clear right away that the public liked our response. Three days later, the same people who did the first customer survey went back and asked the identical group how many had heard about our reaction. The number was 53 percent. When they were asked if they approved of our response, 67 percent said they did. The survey people were astounded at how fast this flip-flop took place simply because I'd stood up and declared, "I screwed up."

The world at large knows that testing is a good idea. The only thing worse than the consequences of testing is the consequences of not testing. Boy, would the complaints pour in then! After all, refrigerators and TV sets are run for hours by the manufacturers before they're boxed and shipped to the stores.

It's the way we went about testing that was dumb. Admittedly, we shouldn't have been driving cars with the odometers disconnected. It would have been so easy to connect them and put on a sticker that said: THIS CAR HAS BEEN DRIVEN BY LEE IACOCCA. IT'S GOT TWENTY-ONE MILES ON IT. I even got about fifty letters suggesting we turn this mess into a marketing opportunity. Some of them came up with the bright idea that we put plaques on the cars with the names of the executives who'd driven them and then sell the cars as "Signature Models." And even charge a premium for them!

At one point, one of my guys got so gun-shy that he said, "Maybe we should simply drop the overnight testing program."

When Dick Dauch, our top manufacturing guy, heard that, he went through the roof. He said, "This week, I had a minivan problem in Canada. Four guys drove new vans home and they all found the same thing: If you make a left-hand turn while you have the air conditioning on, the engine stalls. How are you going to find that out by driving the van once around the plant? In those test drives, they don't even switch on the air conditioning. This program is vital."

The saga came to a close with a plea of nolo contendere. That means "I'm not innocent and I'm not guilty." The criminal charges were dismissed against Frank O'Reilly and Allen Scudder, the two managers involved. Both of them and their families had suffered plenty from the adverse publicity, and yet they were two of the hardest-working, most quality-conscious men I've ever met in my life. They were interested in testing those cars to make sure they were all right. But sometimes the most innocent bystanders get caught up in a messy situation. That's not fair, but that's life. Fortunately, they were exonerated.

As it turned out, most people were understanding. In the aftermath of the flap, it seems my image and the company's image among our customers went up. Sometimes the most reassuring thing you can do is simply to admit you're wrong. I got a ton of mail suggesting Richard Nixon and Ronald Reagan could have saved themselves and the country a lot of grief by doing the same thing.

If you own up to your mistakes, you don't suffer as much. But that's a tough lesson to learn. It would have been a lot better not to have made the mistake in the first place. Maybe I should go back and add that as an axiom to my management chapter.

VII
I'LL SEE YOU IN COURT

When I was about five years old, my father ran into a horse with his Model T. Actually, the horse reared and backed into his car, which was parked. Fortunately, the policeman who was sitting on the horse wasn't hurt. The horse didn't make out as well. They had to shoot him. My father was taken to court and had to pay for the animal.

I still don't know whether it was his fault or the cop's—or the horse's—because the subject was never discussed in our house. It was absolutely taboo. Going to court in those days brought shame to the family, especially for a young immigrant trying to build a new life in America. For as long as he lived my father never talked about the horse, and nobody ever brought it up. It remained a deep, dark family secret.

Times have changed. Today, unless you've committed a crime, going to court is nothing to be ashamed of. In fact, suing somebody or getting sued carries a certain status with it. "I'm suing the son of a bitch" is one way to let everybody know you can't be pushed around. Even getting sued is practically a badge of honor in a perverse way. If you've never been taken to court, you begin to feel that maybe you're not aggressive enough or important enough or rich enough for anybody to bother. (Frankly, I've never even thought of suing anybody in my life.)

The United States is by far the most litigious country on earth. We sue one another at the drop of a hat. About 90 percent of all the civil actions tried before juries in the whole world are tried in the United States, and we're spending about $30 billion a year just suing each other. In the old days, if a neighbor's apples fell into your yard, you worked it out over the back fence or picked them up and made pies. Today, you sue.

I don't know if the problem is too many greedy lawyers or too many greedy clients. Dick Goodyear used to remind me that behind every lawsuit there's a client; the suits don't just pop up like mushrooms, he said. But I kept remembering the old story about the small town with one lawyer, who was starving. Then another lawyer moved to town, and they both got rich. There's more than a grain of truth to that.

When something tragic happens in our complex, high-tech society, it's getting increasingly difficult to decide who was at fault. Sometimes literally dozens of people might share the blame. In my town the worst tragedy to take place in years was the crash of Northwest Flight 255 at Detroit Metropolitan Airport on August 16, 1987, killing 156 people. Within a day or so, some lawyer in Phoenix had an ad in the paper soliciting business from the families of victims; you'd think the guy would at least have had the decency to wait until the bodies were identified. But the biggest louse was a mysterious "Father Irish," a man dressed as a priest who was supposedly passing through town the day after the crash and kindly offered his services to the grieving families. A few weeks later, it turned out that "Father Irish" was a phony. He was actually a shill for a Florida attorney.

Nothing draws the vultures quicker than an air crash. For a lawyer, that's a no-brainer. Somebody is going to pay. For a long while, no one was sure what caused the Detroit crash, whether the pilots forgot to put the flaps down for takeoff, or the air traffic controller put them on the wrong runway, or an engine exploded, or a wind shear caught the plane. In cases like that, it's tough to establish fault. But regardless of fault, you know the victims' families are going to be compensated. And, of course, so are the lawyers.

An air crash is one thing, but we operate under the curious notion that anytime something bad happens to us, somebody else is at fault—and somebody has to pay. It may turn out that a lot of people were at fault, but you can bet that the lawyers will home in on the guy with the deepest pockets, because it doesn't do them a hell of a lot of good to sue somebody who can't pay.

Nobody argues that the families of airline crash victims shouldn't be compensated, or that gross negligence shouldn't be punished. But we've gone way beyond that. The courts were created to distribute justice, but now I'm afraid they're also being used as a way to redistribute wealth. They've become part of our "casino society," and the jackpots have gone out of sight. The court that told Texaco to pay Pennzoil $10.3 billion not only set a world record for a civil suit but also raised the stakes so high that a lot of people had to be thinking: *How can I get a piece of the action?* A $2 billion lawyer's fee would make anybody drool. And there are enough crazy things going on in the courts today to make it worthwhile to throw the dice. Who knows, you might just get lucky.

A couple of years ago, a burglar who was breaking into a school fell through a skylight and successfully sued the school for damages.

In another case, a bank robber was running down the street with the loot when a small explosive device a teller had put in with the bills went off and burned the guy's balls. He not only got caught; he sued the teller.

A man attempted suicide by jumping in front of a train but messed it up, wound up with a couple of bruises, and then sued the transit company because the train didn't stop fast enough to prevent his injuries.

A big drug company lost a case and had to pay $3 million for not posting a warning label that the Food and Drug Administration had specifically prohibited it from using.

My company was once sued by a policeman who was shot in one of our cars; he said the rearview mirror must have been defective because he couldn't see the guy sneaking up on him.

I know these sound like atrocities, but damned if they aren't getting a little commonplace.

Lawsuits have become such a big business that you can even buy shares in them. The *San Francisco Chronicle* ran an ad that read: "Joint venture partners wanted to finance 'sure thing' six million dollar lawsuit against major SF bank. . . . Micro-Vest did it to Computerland, now you can do it to the bank."

That's the same town where a woman once sued the cable-car company after an accident that, she claimed, was so traumatic that it turned her into a raging nymphomaniac. I still remember the headline on the story: STREETCAR NAMED DESIRE.

We can laugh at some of these, but the trends behind them aren't funny. Between 1975 and 1985, there was a 1,000 percent increase in product-liability cases in the federal courts, an 835 percent increase in malpractice awards, and a 401 percent increase in liability awards. The courts are jammed, and people with legitimate cases have to wait in line for four years or more to be heard.

The impact of all these suits goes beyond the courtroom, however. The fear of getting sued has changed the way we relate to one another, the way we do business, and even the way doctors practice medicine. This epidemic of lawsuits is killing the doctors. I feel sorry for them. Every patient is a potential plaintiff these days, and that has to throw cold water on the old doctor-patient relationship. We all pay through the nose for the extra tests and procedures doctors do just to be sure their patients won't take them to court for not doing enough. Defensive medicine is one of the main reasons health care costs have gone up faster than just about any other cost—in fact, they've risen at more than double the inflation rate, every year.

I'm always asking our people to tell me all the costs that go into a car, because identifying them is the only way to get them out. Well, employee health coverage adds about $600 to the price of every car Chrysler makes, and we estimate that 10 percent of that is wasted money directly related to defensive medicine. It pays for unnecessary diagnostic tests, hospital admissions made "just to be safe," and, of course, the doctor's malpractice insurance premiums.

You won't find it on the window sticker, but $60 of the price of one of our cars is a medical liability cost. The amazing thing is, the medical liability costs are higher than the product liability costs on the car itself. In other words, it costs Chrysler more to try to keep our employees' doctors out of court than it does to try to keep ourselves out of court.

Some of the doctors are fighting back against the lawyers. In one small Midwestern town, an obstetrician refused to deliver a lawyer's baby because of a malpractice squabble. Now I think that's going too far. He might have been just as effective had he agreed to deliver the kid, but only on condition that the father sign a pledge that the kid would never grow up to become a lawyer.

Occupational birth control is a good idea, because we've got a population explosion in the legal profession. Back in the 1950s and '60s, the number of lawyers grew at about 2 percent a year, then it doubled to 4 percent in the '70s, and now they're reproducing at the rate of 8 percent every year. So it's no coincidence that the number of court cases goes up by 10 percent a year. All these people have to eat! Lawyers are supposed to help resolve conflict, but you don't have to be an absolute cynic to see that they have a vested interest in promoting it.

By now you're thinking I'm a lawyer basher. But I'm not. I find it hard to live with them, but I find it just as hard to live without them. What worries me is that lawyers are all a little more equal than the rest of us. Everybody in the country has the same rights, but lawyers are trained to use those rights. They go to school to learn how to push those rights to the fullest.

If you're not a lawyer, it can cost you a lot of money to insist on what you're entitled to. I found out when I was in graduate school at Princeton that you can't go it alone. One day I parked my car in a two-hour zone on Nassau Street, went to a two-hour lab, and came back to find a parking ticket stuck on my windshield.

When you're young and feisty, you don't stand for things like that, so I went to court to contest the ticket. I told the judge that I knew I had parked for less than two hours because I'd been late for the lab and caught hell from my professor; class was dismissed five minutes early and I went right to my car.

The judge was a nice old guy but after I'd told my story he asked me if I had any witnesses to back it up. Of course, if I'd hired a lawyer I could have beaten the rap. I thought of subpoenaing my professor to swear that I was late for class, but he was already pissed off at me so that didn't seem too smart. Being a twenty-one-year-old idealist, I kept saying to myself: *It isn't the lousy ten dollars—it's the principle of the thing.*

But suddenly the price of justice started looking too steep. I was too busy trying to get a B in advanced thermodynamics to waste my time in court. So I paid the ten bucks and dropped the whole thing, but I learned a valuable lesson about all those rights we have under the Constitution: Sometimes the price of exercising those rights is too high. And I'm sure that's why so many civil cases today are settled out of court. Caving in is just a lot less expensive than fighting.

Lawyers have a corner on the justice market. If you want justice, it isn't free; you have to hire a lawyer to get it for you. And if you get mad at lawyers, you'd better hold your tongue, because the ultimate irony in all this is that the only person who can defend you against a lawyer is another lawyer. So the problem feeds on itself. And that's why companies like Chrysler hire platoons of lawyers—to protect themselves from all the lawyers they don't hire.

Some of our lawyers at Chrysler work full time just to see that we comply with all the laws and regulations that apply to us: environmental laws, safety laws, tax laws, stock market regulations, even local zoning laws. We have lawyers review our advertising before it runs and our press releases before they go out. Before we name a new car, the lawyers have to make sure nobody else has prior rights to the name.

Not only that. Our chief financial officer is a lawyer. Our labor relations vice-president is a lawyer. The head of our tax department is a lawyer, and so is the man who directs our state and local government affairs and the man who deals with our securities analysts. We wouldn't dream of holding a board meeting without our chief legal counsel present. And then, of course, we have other lawyers to defend us against the two thousand or so lawsuits that are filed against us for one reason or another every year.

Altogether, Chrysler has about a hundred lawyers on the payroll —more than the number of stylists or interior designers we employ— and that doesn't include the hundreds of outside lawyers we have to hire every year. Believe me, there's nothing unusual about Chrysler. We're typical of every big company.

If he's not careful, a CEO can wake up one morning and find the lawyers running the company. With all the statutes and regulations on the books and all the people on the outside just waiting for the slightest excuse to sue, it's easy to get paranoid and call in the lawyers to make all the big decisions. The law is often ambiguous, and the easiest way to stay out of trouble is not to do anything.

Of course, that's also a good way to go broke. That's why I look for corporate attorneys who are also solid businessmen. I can't afford to have lawyers around me who simply tell me every day what I can't do. I want lawyers who tell me how to do something and stay within the law, but who still have enough moxie to stand their ground when they have to.

It's crazy. There's not a single clause in our corporate plan about how to break the law or harm somebody. In fact, we hold our people's feet to the fire to see that they obey the law to the letter. And yet we budget millions in advance every year for legal costs. It's become just another cost of doing business, and like every other cost, it's passed along to the consumer.

Unfortunately, some of our biggest competitors today, the imports, don't have those same costs to pass on to their buyers. In other countries, people aren't as quick to sue. The auto companies there don't need to employ the platoons of lawyers that we do, which gives them a competitive advantage over us in the marketplace.

So far, this obsession with litigation seems to be pretty much confined to the United States. But while we're wasting all this time and energy suing one another, the other guys are concentrating on beating our brains out. They know that the real Mr. Deep Pockets is the American consumer, and so they don't spend their time in court; they spend it keeping costs down to get at him.

The Japanese have about as many lawyers as we have sumo wrestlers. They don't need them, because they rarely sue each other. There is only one lawyer for every 9,600 people in Japan, compared to one for every 360 in the United States. On the other hand, they have one engineer or scientist for every twenty-five people, while we have one for every hundred. That tells you the whole story about screwed-up priorities.

But we could be in for a break. I gave the keynote address to the American Bar Association just a few months after the Japanese announced that American lawyers could now practice in Japan. They lifted the ban on April Fool's Day 1987, and I told the ABA delegates to get their visas ready because here was an opportunity to serve their country. I encouraged them all to go to Japan and hang out a shingle. It would do wonders to reverse our trade imbalance if only we could get the Japanese tied up in court as much as our companies are.

In particular, I told the lawyers to see if the Japanese would buy the concept of "punitive damages," because that's raising hell here and yet it's something the Japanese and most other countries have never heard of. In America, if you lose a suit, first you pay the real damages to the defendant and then you're liable to get hit with punitive damages as well. They've not only sent the cost of getting sued sky high, but they've also added a lot more risk to doing business here.

Punitive damages are intended to punish gross negligence and reckless disregard for somebody else's safety or rights. They used to be rare, but they aren't anymore. In fact, they're so common that I'm afraid we're beginning to punish more than gross negligence in this country; we're beginning to punish the normal risks that simply can't be avoided when you produce almost any kind of product.

Another dumb thing about our system is the way we pay lawyers. Under our contingent-fee arrangements, anyone can hire a lawyer and file a suit without coughing up any real money. The lawyer collects only if he wins the case. But that can be one helluva incentive because he often takes a 30 or 40 percent cut out of the award.

In most other countries, this system is illegal. Lawyers must be paid up front, meaning you'd better make sure your case is solid before you proceed. In fact, in places like Brazil, you'd really better have a terrific case before you head for court, because if you lose, not only do you have to pay your own lawyer, but you have to pay the opposing attorney anywhere from 10 to 20 percent of the amount you sued for. Guess what? Brazil has a lot fewer lawsuits than we do—and the damages asked are a lot less absurd.

Now if we don't make some changes, but continue to penalize people for taking normal risks, then we're also going to end up punishing progress and punishing competition. Most drug companies, for example, don't want to have anything to do with vaccines—they're too risky. A simple shot for diphtheria and whooping cough that cost 12 cents in 1980 costs $12 today, and almost all the increase goes into a liability fund. We've got some awful new diseases showing up lately that seem to be crying out for a vaccine, but you can't blame the drug companies if they prefer to spend their R & D funds on something a little safer—like growing hair.

There used to be eighteen companies making football helmets in this country, but the liability crisis has pared them down to two. Nobody makes gymnastics equipment or hockey equipment in the United States anymore because it's too risky. A small company in Virginia that used to produce driving aids for handicapped people folded because it couldn't afford the liability insurance. We've almost stopped manufacturing light aircraft in this country because the liability insurance is more than the cost of production.

The stories go on and on, and if they keep up, I'm afraid one day we're going to say the hell with it, it just ain't worth it, and pack it in.

Every business and every product carries risks. I'm more sensitive to it than most because I'm in the automotive business, and cars and trucks are involved in about half the tort cases in the country. Get this: Cars get safer every year, but the number of lawsuits continues to go up.

The only perfectly safe cars I've ever seen have been in museums. And as long as we have to put a human being behind the wheel, we'll never eliminate traffic accidents. For years, there was a tree in the middle of the Sahara desert. It was a freak of nature, the only tree within hundreds of miles in any direction. One day during World War II an Englishman ran into it in a tank and got hurt. If it happened today in this country, he'd sue. I don't know who, but I guarantee that his lawyer would find somebody to sue.

We don't live in a risk-free world, and we never will. If we continue to punish risk, we're bound to fall behind the countries that don't. Which doesn't mean you can throw caution to the wind. When it comes to nuclear power, for example, I want the risks reduced to the absolute minimum. But our idea of minimizing risks is to stop building nuclear power plants, largely because the legal challenges have gotten too expensive. If a totally safe plant is ever going to be built, it probably won't be here, because other countries have their engineers working on the problem while we've turned it over to the lawyers.

Don't forget, it's not the lawyers who build safe nuclear plants, or develop vaccines, or come up with a better braking system for an automobile. They are simply professional advocates. They get paid to take sides, because when we get in trouble we want the meanest bulldog we can find sitting next to us in court; we don't want a potted plant.

That kind of advocacy is fine, and it's necessary at times to settle our differences when everything else fails. But the unbridled advocacy we see today is creating differences we didn't even know we had.

Unbridled advocacy is like unbridled capitalism—too much of a good thing. We don't have pure virgin capitalism in this country anymore. We came close to it around the turn of the century, until we woke up one morning and discovered that a couple of guys owned all the steel, and a couple more owned all the oil, and a few others had grabbed all the railroads. What we found out is that if you allow yourself unrestrained competition in the marketplace, pretty soon you wind up with no competition at all.

I'm afraid that all this litigation is taking us back to the Robber Baron days. The Constitution gives us the right to use the courts, which makes us luckier than most of the people in the world. But whenever you start to abuse a right, you risk losing it. If every one of us keeps pushing every one of our rights as far as we can, then one of these days we'll have no rights left at all.

We can't afford to let the lawyers and the courts settle every conflict that comes up. Nothing is more fundamental to the success of a pluralistic society like that of the United States than the honorable old art of compromise. And nothing tells me we're losing it more than this epidemic of litigation.

Going to court ought to be the last resort, but it's becoming the first resort. "I'll see you in court" is the modern-day equivalent of "I'll meet you in front of the Last Chance Saloon at high noon."

And once a suit is filed, it's a gunfight. There's no middle ground, just win or lose. The guy who in real life is willing to meet anybody halfway suddenly turns into a tiger.

In that kind of environment, among the first casualties are truth and candor. Confession is good for the soul, and when you offend someone, even unintentionally, it feels good to say "I'm sorry." But when there's a chance that you might end up in court, you'd better think twice.

The sad fact is, being forthright and honest can set you up for those who aren't. Watergate gave stonewalling a bad name, but whether you're guilty or innocent, the first advice you often get from a lawyer is "clam up."

That's what we did at Ford in the late '70s when we were bombarded with suits over the Pinto, which was involved in a lot of gas tank fires. The suits might have bankrupted the company, so we kept our mouths shut for fear of saying anything that just one jury might have construed as an admission of guilt. Winning in court was our top priority; nothing else mattered. And, of course, our silence added to all the suspicions people had about us and the car.

I learned something from that. The Pinto was a legal problem and a public relations problem, and we chose to deal only with the legal problem. Ever since then, whenever we've had a similar choice to make at Chrysler, we've done the exact opposite. We've chosen to look past the legal consequences and go public with the whole truth.

See, you do learn from experience. Reminds me of that old Pennsylvania Dutch saying I used to hear so often while growing up: "Why do we get so soon oldt and so late shmart?"

VIII
THE PRESS — WE CAN'T LIVE WITHOUT IT

By now, the expression "Don't believe what you read in the newspapers" has become a cliché.

Or, as one of our national politicians put it after another day of hard knocks from reporters: "The man who reads nothing at all is better educated than the man who reads nothing but newspapers." He also recommended that an editor "divide his paper into four chapters, heading the 1st Truths, 2d Probabilities, 3d Possibilities, 4th Lies. The first chapter would be very short."

We all know who said those things, don't we? Richard Nixon, right? Or was it Gary Hart?

Nope. Thomas Jefferson.

It just goes to show you, some things never change. In the nineteenth century one etiquette book advised the upper crust that it was bad form to order a newspaperman kicked down the stairs simply because he was earning a living in a most disagreeable way. Today you may not find that sort of advice in Ann Landers's column, but much of the public still looks at reporters in about the same way they do used-car salesmen.

Yet no matter how you feel about the press, there's no question that its role is vital to a free society. In fact, in today's global economy and world of instant communications, what happens in Timbuktu or Lapland could affect our lives too. I'll bet you'd never heard of Chernobyl—I hadn't—and then suddenly you were squirming every time you watched the evening news in your living room. So, unless you've got a lot of free time on your hands and have won a Frequent Flier Grand Prize, the only way to know the score is through the printed word and the infamous boob tube.

The power of the press is astounding. A couple of years ago in Washington, I made a wisecrack at a press conference on some inside dope I had that the yen, which was at 180, would soon drop to 150. Actually, Jim Baker, the Secretary of the Treasury, had yelled at me in the hall that 150 looked like a nice, round number to shoot for, so the next day I quoted him.

All the reporters laughed—all but one. He put it on the wire. This was about eleven o'clock in the morning. Before the end of the afternoon, the yen had reached a new postwar high against the dollar, trading had been stopped in all currency futures, and the bond market had dropped two points. And the market blamed the whole damn thing on me!

I suspected for years that somebody was rigging the exchange rates. I just never knew it was so easy. (If I had, I'd have started making wisecracks a long time ago, believe me.)

Nobody understands the power of the press better than politicians, because reporters have as much (or more) to do with who goes to Washington as do the people who pull the levers in the voting booths. Look at the 1988 Presidential race. Gary Hart figured he could have a few little adventures right under the press's nose and get away with it. Fat chance. He got a fast trip back to Denver (although apparently he hadn't had enough abuse, because he came back for more). Joe Biden thought nobody would notice if he borrowed a couple of good lines to add some zap to his speeches. *The New York Times*, the *Des Moines Register*, and NBC News zapped him instead. The press even got credit for a gag: "Did you know Biden's writing his autobiography? It's called *Iacocca*." A prying press forced the two reverends (of all people), Pat Robertson and Jesse Jackson, to come up with some quick explanations of why their wives' pregnancies began just a tad before their marriages did. By now, I'd bet that any politician with a black book has deep-sixed it in the Potomac.

The press notices everything today, and with the immediacy of television, lets everyone else in on the scoop. When newspapermen camp outside politicians' homes with telescopes, it's clear that reporters have reached a new level of snoopiness.

It's small wonder, then, that everyone in Washington is caught up in the world of image and perception. Every ten seconds the politicians get the flashes on a TV screen: How did they cover my speech? What did they say about what I had for lunch? Sometimes the powers in government seem more interested in whether they get two or three lines in the newspaper, or thirty seconds on TV, than in how well they do their jobs. Public officials, you may have noticed, rarely do anything after five P.M. because whatever they say won't make the six o'clock news. And if it's not on the tube, that's as if it never happened.

Even though they could be pretty ornery, many members of the press corps were once in awe of politicians. It wasn't that long ago that FDR's "Giggle Gang" of reporters used to station themselves in the front row at the President's press appearances and laugh uproariously at his jokes, even the bad ones. That was one way to avoid being banished to Roosevelt's Dunce Cap Club—for hapless reporters who irritated the President. Now the situation's flip-flopped. The politicians make sure they laugh at the reporters' jokes, and it's the reporters who distribute the dunce caps.

Politicians are willing to make total fools of themselves if it wins them some brownie points with the press. A few years ago Nancy Reagan showed just how important the press had gotten when she put on a skit at the annual Gridiron Club dinner for the Washington press corps, to try to spruce up her bad image. She had increasingly been depicted in the media as a rich snob, and so she got dressed in Salvation Army shlock and sang a song called "Secondhand Clothes" to the tune of "Secondhand Rose." I have a hard time imagining

Harry Truman showing up to do a little skit about the A-bomb because it might help improve his image.

In Washington they say you've arrived if the bigwigs poke fun at you at a Gridiron dinner. Well, I've arrived. One skit had a punch line: "You don't need an airbag in your car. You've already got one in Iacocca."

Teddy Kennedy quipped that he heard I might decide to become a presidential candidate, except that I was baffled about which party to join. "Anyone who would take $1.2 billion from the government ought to be a Democrat," he said, "and anyone who would pay it back is dumb enough to be a Republican."

Even the President let me have it. He said I would like to be President, but only if the horns on all the White House pool cars played "Hail to the Chief."

I do have to wonder what the press is really after. Is it truly trying to help elect the best possible leaders? In 1985, I attended a background luncheon with editors and reporters of *The Washington Post*. I've been doing that for twenty years now, and I always start off by kidding Kay Graham, the *Post*'s chairman, about her Mercedes and telling her that I'm going to sell her an American car before I die. Then I needle Ben Bradlee, the executive editor, who majored in Greek, that most of the stuff written in his paper comes off like Greek to me. After that, I'm the one who gets worked over.

In the course of the luncheon, the *Post* people asked me why I didn't run for President. I listed my usual reasons.

"Well, assuming you're not running," David Broder asked, "and assuming a businessman would be good for the country, is there any Fortune 500 head who could be a candidate for President?"

I thought about it but couldn't come up with a quick response. Two years later, I'm still pondering the question.

Then another reporter bored in on me. "Of all the potential candidates, then, who could be elected?" They didn't ask me a thing about who would make a great President, just who might pull off the nomination.

"On that basis," I said, "it would be Bob Dole versus Mario Cuomo." (Of course, I said that back in early 1985.)

I know it sounds cynical, but I really believe the only way you can be a true contender in this day and age is to be a good debater and have the TV tube portray you as sincere while you're debating. That's just the nature of the beast.

Sometimes you can be skeptical of the motives of the press, but anyone who underestimates its power is naïve—no, plain stupid.

For years I've contended that the two most powerful groups in the country—other than TV anchormen Dan Rather, Tom Brokaw, and Peter Jennings—are the administrative assistants in the Congress and the country's newspaper editors.

The editors set the agenda of the nation. They decide what's page one, what's page thirty-eight, and what's no page at all. Now that's clout.

Media power—from nine thousand miles away in America—got Marcos a one-way ticket to Hawaii. (And cost the missus five hundred black bras and three thousand pairs of shoes.)

In 1985, when Mikhail Gorbachev became the new leader in Russia, he was immediately hailed all over the world as a PR whiz. (Probably because he doesn't eat with his fingers and pound his shoe on the table.) Well, the next year, when he forgot to mention to the Swedes and a few others that there was this little cloud heading their way from Chernobyl, he wasn't a media hero anymore; he was a media bum. Now that he's letting dissidents leave the country and touting entrepreneurship, he's been elevated to hero again. Fact is, it's getting tough to stay current. If you go away for a long weekend and skip the papers, you might come back and find out that your worst enemy is now your best friend.

It's scary, the power of the media today. (Not bad or good, necessarily, but scary.) I wonder if we know yet how to use it wisely. When the Russian reactor blew up, one commentator said about technology: "Our knowledge may exceed our understanding." The same thing may be true of mass communication today.

As Pope John Paul II told an audience of media leaders during his visit to the United States last year: "In a sense, the world is at your mercy."

Because the press has become a far more powerful institution than it was, it has to be careful how it uses its muscle. If the press wants all the rights it's been handed—and it's been handed a fistful—then it had better practice certain codes of behavior.

Which brings me back to my old line: Life is nothing more than accountability. Kids are accountable to their parents (or should be). Workers are accountable to their bosses. I'm accountable to my board. But I'm not too sure who the press is accountable to. It likes to picture itself as the watchdog of society. But who's watching the watchdog?

To some extent, it's inevitable that a lot of people distrust the media. The press, after all, often brings us the news we don't particularly want to hear: airplane crashes, murders, sex scandals—or more minor disasters, like car recalls.

But the problem goes beyond that. Such an adversarial role has developed between the press and many of the people and institutions it covers that the two sides hardly ever trust each other. Part of the problem, I'm convinced, is that some apple-cheeked, eager-beaver reporters are so intent on making a name for themselves that they play fast and loose with the facts. The best (or worst) example of this is Janet Cooke, the young *Washington Post* reporter who several years ago fabricated an entire story about an eight-year-old heroin addict, thinking it would win her a promotion. Well, it did more than that. It won her a Pulitzer Prize, until some poking around by other reporters uncovered the fraud. She lost the prize and her job.

Too many reporters, moreover, have a tendency to see the worst in everything. Not long ago, the *Kansas City Times* alleged in a series on athletic recruitment that the mother of a Wichita State University basketball player had gotten a new car and a house as a payoff for her son's slam-dunk abilities. Turned out (as another paper, the *Wichita Eagle-Beacon*, discovered) that the money for the purchases actually came from the settlement of a medical malpractice suit.

The press is quick to announce when someone has been accused of something, but slow to report when that same person has been

Ten years at Chrysler, before and after. The eyes and the hair go first; the brain, later.

Focal Point of Farmington, Michigan

Lost two daughters, gained two sons.

Bachrach

The Statue kept me—and the cartoonists—busy.

A gift from Kathi:
Even at six months—a chip off the old block.

Ia-*cocoa*—my mug in chocolate.

I put my stamp on the Caribbean.

If the car business ever goes under, have I got a wine cellar!

On the set of *Miami Vice*: Don Johnson, eat your heart out!

UPI/Bettmann Newsphotos

Iacocca (sired by Nick Iacocca) talks to Iacocca (sired by Seattle Slew).

I liked the car so much, I bought the company.

The book had an impact—at home... and abroad.

"Gentlemen, let us open to page 82 of our Iacoccas."

Even honest mistakes can kill you.

IACOCCA / NUNN
IACOCCA '88
IACOCCA FOR PRESIDENT

I never made it to New Hampshire.

Michael Schwartz/*New York Post*

All Lee needs is a line.

Write Iacocca's campaign slogan and win a weekend for two in Washington

Madison Avenue

Draper Hill, © 1986, *The Detroit News*

I hope so too!

Home of the Iacocca Institute. BMOC: Big Man On Campus—finally!

Lehigh University, photo by James P. Meleta

Me and my hero, Joltin' Joe, fifty years later.

Paul Lester Studios

My real hero, Mom, talking straight to me.

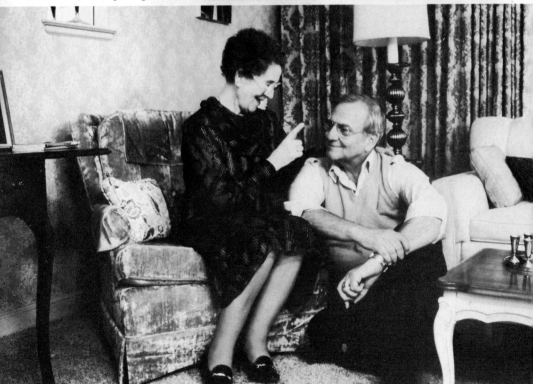

Harry Benson

cleared of the charges. Indictments always seem to get reported on page one. News of their being dropped, on the other hand, has a way of turning up on page sixty-two, just before the want ads.

Sometimes, the press seems to have a preconceived idea of what it wants to say, and then goes out looking for someone to say it. I felt that I was recently an unwitting victim in an episode I refer to as the Connie Chung Ambush.

In April 1986, I called a press conference in Washington to show off our new Dakota truck. The Dakota was a super product, America's first midsize truck. It cost $600 million to bring to market, boasted unbeatable performance, and it had a paint job better than a Mercedes. We were damn proud of it.

After the conference ended, one of my press guys walked up to me and said, "Listen, Connie Chung of NBC is here, and she wants to ask you about truck safety."

It was a bit unusual for this kind of press conference, but because I had met her once at Tom Brokaw's house, I went over and asked her: "What's up?"

She told me she had to have a little piece for a new show called *1986*.

"All right," I said. "What kind of questions have you got?"

We walked over to our truck, and suddenly she introduced me to her director, and the crew began pinning a mike on me. Christ, that was a little much. For a minute I thought they'd call for the makeup man. As I was pondering how I'd gotten hooked into this, Connie Chung started interrogating me about why the new Dakota didn't have headrests and why there were no side beams in the doors.

I had no idea what was bugging her, but I did my best to answer her questions. I told her that, first of all, Chrysler complied with all government standards for truck safety, and second of all, some of the safety features we put on cars are not relevant to trucks. There was no evidence, for instance, to suggest that headrests in a truck would prevent serious injuries or even more minor ones such as whiplash; in fact, there's precious little evidence that they do much good in cars, beyond cutting down on whiplash. What's more, I said, truck safety statistics are generally better than cars', for a simple reason: the second law of physics. Trucks are heavy. If a four-thousand-pound truck hits a two-thousand-pound car, you'd better believe the truck will win every time.

What was Chung's reaction? Aren't you saying, she told me, that you wait for the government to issue rules rather than promoting safety as the right thing to do on your own? I told her that was nonsense—but only because the word *bullshit* wouldn't play on the air.

When the show was broadcast, I found out why I was being questioned so persistently about headrests. The producers had footage about a guy who had gotten brain damage from a truck accident, and my interview was juxtaposed with a flaming crash and a guy on a stretcher! Leaving the distinct impression that here's this injured Joe and here are these heartless tycoons in Detroit who don't give a hoot about him.

To make matters worse, there was nobody from Ford or GM on the show. I was the anointed industry spokesman—and hangman.

A couple of years ago, General Motors put in force a regulation forbidding its management from appearing on a documentary unless they were afforded the right to edit it. As a result, nobody at GM ever goes on any documentary, which is just dandy with GM. I certainly don't believe in that kind of policy, but another couple of Chung episodes could change my mind in a hurry.

Even when you get put through the meat grinder, you can't get overly paranoid about the press, or you'll be too scared ever to sit for an interview. And not talking to the press, as Richard Nixon found out, can be a lot worse than talking.

I have a few simple rules that I try to follow when I deal with reporters. I think these rules help me get fair treatment—and help the press get their story.

Probably the most important thing I do is to be accessible, and I don't mean accessible at my convenience. Most CEOs like to hold press conferences when they're reporting the best quarter in the history of mankind. When they have a dismal quarter, they make sure they're visiting the Helsinki plant.

There are plenty of times when I bring in the press and say: "I usually call you together when I have happy news; today I'd rather be in Philadelphia. Because the news is rotten."

My unwavering rule is to hold a press conference to announce every quarterly result—whether it's up, down, or sideways. That's the only way to be fair to reporters. They've got a story to write no matter how our earnings turn out—their editors don't give them the day off if we have a dud of a quarter—so why shouldn't I be available to explain why we fouled up?

I've got another rule I'll bet nobody else has thought of. It's one of my deepest, darkest secrets: I tell the truth.

I'm convinced that if you level with the press, your coverage will be fair and balanced. There's plenty of good and bad in every company. So just share the whole mishmash with them. Believe me, they'll find out the bad stuff anyway. Why not take them behind the scenes and save everyone's time?

It's suicidal to hold your cards close to your vest. The government always likes to keep something back by invoking the principle of national security. How much of it is really top secret I sometimes wonder. In the business world, we often tell reporters we can't reveal something because we don't want our competition to know about it. Well, I think plenty of corporations badly abuse that privilege, and the press wises up to that excuse fast.

Reporters want some balance; our job is to help them get it by supplying the facts. Just tell them direct: "You're going to say some good things about me, which is perfectly fine with me. But there is some bad stuff, too, and rather than your spending two weeks nosing around trying to find it, let me save you a lot of trouble and hand it over."

A lot of corporation heads get one negative line in one story and immediately cancel their subscription. Then they shoot their PR guy. But they don't bother to look at the 412 good lines. Because they run a big company, their attitude is: "How can that little whippersnapper of a reporter come in here and criticize me?"

If I had to grade the business community on its cooperation with the press, on a scale of one to ten, it would rate about a two. When reporters call for interviews, most companies tell them to climb a tree. Then if they get bad press, they're the first to complain. When I was at Ford, in fact, some of our top executives often considered winning over the media by becoming part of them. They seriously weighed the possibility of either buying a big newspaper or a network like ABC. Then, in their naïveté, they thought they could impose their philosophy on the world and never have to worry a whit about negative press again.

Although I strongly advocate being forthright with reporters, there are times when you really can't talk about something because it's genuinely a trade secret, at least for a while. A reporter may say to me: "Hey, I hear you're having problems with the start-up of your '88 cars. What's going on?"

Instead of just answering "No comment" or "Why don't you shut up?" I'll tell him: "Well, in time I'll be able to fill you in. But now's not the time."

You have to know when to pass. I frequently say, "I'll see you next month on that. I'll be ready then, but I'm not today." If the reporters know from your past behavior that you really will give them the facts next month, they won't get all riled up.

Which is not to say that some reporters won't keep hammering away at you. President Reagan has always gotten annoyed at Sam Donaldson of ABC. One day the President even asked me: "How do you think I should handle him?"

"I guess I'd try to banter with him," I replied. "But you have to remember that because he has the microphone, he's always going to get the last shot, just like a comedian on the stage. The only weapon you have is to ignore him. Call on NBC and CBS, and then conveniently run out of time when ABC's turn comes up. After a while, the guy will get the message."

I play it that way myself. If someone's a royal pain in the ass, I'll let the rest of the guys ask all the questions. After all, it's my press conference.

* * *

A lot of reporters like to tape interviews these days. I've had reporters come in, tape me, and then just print the transcript verbatim—sentence fragments, dangling participles, and all. If I wanted my transcript released, I'd buy some radio time and just go on the air myself with it. So I've established a rule on the use of tape recorders: Unless I know a reporter, I won't let him or her record me.

Once I do allow a reporter to tape me, I make it clear that he can't build up his archives with a body of tape to be used any time he wants. The information is for this particular story—and that's it.

Not long ago Henry Kissinger gave an interview for a book and, much to his chagrin, discovered the material in *Penthouse* magazine between a photo spread of a nude woman and an article on wrestling. He promptly sued.

I had a similar experience. *Life* magazine wanted to do an article on me and the Statue of Liberty campaign. They commissioned a young free-lance writer named Bob Spitz to interview me. He brought along a tape recorder, and even though I had never laid eyes on him, I allowed him to tape me because *Life* vouched for him. What I didn't know as we travelled around the country was that his mike was always on. He got lots of outtakes.

Eventually the *Life* piece came out. There was precious little on the Statue of Liberty. Most of it was about my personal life. But that's not what gagged me. After a few months the latest issue of the magazine *New Look* appeared on the newsstands, and lo and behold, it featured an exclusive interview with Lee Iacocca by Bob Spitz. Like *Penthouse*, *New Look* was a girlie magazine. There I was with a bunch of girls wearing a lot less than we put on them for a Dodge commercial. As an experienced marketing hand, one thing I know for sure is you never want to get featured in the press between two nude girls.

The piece was arranged in a question-and-answer format—but I didn't recognize too many of the answers. I was almost tempted to sue, except that I'm supposed to be doing my best to keep down all the litigation in this country.

There was one final justice. The issue of *New Look* with my interview turned out to be the last. The magazine folded.

* * *

Just as the press expects certain things from me, I expect a few things from the press. Number one is that reporters do their homework before they arrive. When I give a speech at the Waldorf, I've got to know my subject better than anyone in that room. Similarly, when a reporter comes in for an interview, he ought to know his stuff. When some of these guys waltz into my office, I really wonder where their editors found them. They're covering the auto business, and maybe they know how to drive a car. I wouldn't lay a bet that they could find the spare tire.

I'd also be a lot happier if the press did a little more checking around before making a big splash out of something. And I'd be thrilled by reporters who are willing to believe what people tell them, at least some of the time. Today you can be going along minding your own business and find yourself dragged into a running story of major proportions.

Take this whole thing about me as a candidate for the highest office in the land. It started as a gag in *The Wall Street Journal*. Other reporters read that story and said, "Well, why not? He'd make as good a President as anyone." (After all, at one time or another, Alfred E. Neuman and Pat Paulsen have also been touted.) And so the press took the story and ran with it—for miles.

Suddenly there were articles practically every day about my supposed candidacy. A number of them said I was "hungry" to be President. News to me. I kept telling people I wasn't interested. I even told them why. But the more I said no, the more they told the world I really meant yes.

When I'd come no closer to appointing a campaign manager, the press must have gotten a little bored. It was time to conclude that I'd make an absolutely dreadful President. Every day the papers featured stories asking me why I ever threw my hat in the ring to begin with. As best as I can recollect, I never took it off my head.

When the press isn't writing about my presidential aspirations, it's writing about my health—and in the same wild way. In 1964, I had a little upset stomach and I made the mistake of taking some Mylanta in front of one of my friends. Then he made the mistake of telling a reporter for *Time* magazine about it. Before I knew it, I was characterized in *Time* as a pill-popper. For twenty-five years now, in story after story, as well as in three books, I've been described as a hypochondriac. According to these tall tales, I've spent my whole life taking my temperature, swallowing pills, checking my blood pressure, and hoping against hope I make it through the day.

Well, let me set the record straight. Since that stomachache twenty-five years ago, I've had one episode where my big toe was hurting and the doctor thought I had the gout, so he gave me a pill. Other than that, I've taken a pack of Metamucil every day for twenty years, long before fiber diets became the rage, and that's only because I have a little diverticulitis problem. I also take an aspirin a day, and that's only because fifteen years ago Dr. Christiaan Barnard, the noted heart surgeon, told me it was a good thing and would keep my blood platelets slippery, whatever that means.

Now my old standbys—Metamucil and aspirin—are in the news. It seems that Metamucil helps lower your cholesterol level, and aspirin has magical powers for your heart. If these guys are right, I should live to be a hundred!

Another expectation I have is that the press use some judgment and not simply go with whatever seems to be the most sensational angle of a story. I'll never forget the day before Liberty Weekend. After four years of hard work, I had arrived in New York literally in the nick of time for the press conference to launch the big Fourth of July bash.

Unbeknownst to me, rumors were circulating that we couldn't get insurance for the fireworks. Before I could say one word about all the terrific plans for the weekend, the New York media started bombarding me with questions about whether or not we had obtained insurance for the fireworks display. I didn't even know we had a problem. During the entire conference, the reporters kept howling, "We've got a deadline to meet. Are there or aren't there going to be fireworks?"

I said, "Hold it. We're about to kick off a fabulous celebration. We waited a hundred years for this and you guys want to focus on a last-minute glitch of some kind?" Frustrated, I finally bellowed, "I dare any insurance company in the world to stop this party now. They wouldn't have the guts!" But I couldn't focus those reporters on anything else. Nothing I said could get them interested in the real point of Liberty Weekend.

I knew we would get the insurance, and we did. (I hope you all enjoyed the fireworks; the premium wound up costing us $4.5 million for seventy-two-hour coverage.) Sure, reporters need grist for their mills, but sometimes in their rush to cover a crisis they don't keep their eye on the main event.

On another occasion, I had agreed to give a speech on trade before the House Democratic Caucus, meeting at the Greenbrier resort in West Virginia. All the top hitters of the Democratic party were there, and all the press guys were there too.

Well, I gave a strong speech, one of the strongest ever. I didn't pull any punches. I said that President Reagan ought to warn his good friend Prime Minister Nakasone to start cutting Japan's trade surplus with the United States before relations between the two countries really got mean. I also said the President should call Nakasone on the phone and tell him: "Do it, or the Congress of the United States will do it for both of us! *Sayonara.*"

I must have hit a nerve with House members because I got a standing ovation, and Jim Wright, Dick Gephardt, and John Dingell came up afterward to tell me I was right on.

The next day I picked up the paper and saw that Congressman Bob Matsui, who is of Japanese descent, was calling me a racist for the speech, and especially for my use of *sayonara*. Maybe I'd seen too many old movies, but *sayonara* had always meant a fond farewell to me. And everybody I've asked since says the same thing.

I was furious, maybe as mad as I've ever been. I'd been called a lot of things in my life (some I've even deserved), but nobody had ever called me a racist.

I said to myself: *This is ridiculous. Now nothing about the trade problem is going to make the papers.* And sure enough, all the coverage focused on the racism charge. The key point of my speech, and the whole reason I'd gone to the meeting, had disappeared into the mist.

Now, there's another form of press that's starting to bug me. It's called writing a book. I really had no idea my book would spawn a whole cottage industry. But ever since I wrote my autobiography, no less than six supposedly nonfiction books have appeared to capitalize on the Iacocca owner field out there. They all had one thing in common: They researched my life, more or less, and decided they knew how to revise it—and change my own history a little.

I was particularly fascinated by one guy's rebuttal to a throwaway line I had used in my first book, implying I'd been a bit of a rake in my youth.

Well, this writer actually went back to Allentown and dug up a couple of my high school heartthrobs. Keep in mind that forty-five years had gone by, and these girls were now sixty-three-year-old grandmothers. I think the writer was actually trying to affirm that I had made out in the back seat of my convertible, but alas, all the girls he talked to said I was more like a virgin altar boy. Thank God.

I was amused, because one of my old flames could have told him a cute story, which she had evidently forgotten as time went by. On one of my early dates with her, we had gone to a drive-in for a hamburger and a milkshake, and then we got into some serious necking. After a while she jumped up, straightened her blouse, and yelled, "For Chrissakes, Lido, you buy a girl a lousy twenty-five-cent milkshake and then try to squeeze it out of her." I was so shocked by her sudden outburst I wore mittens on the next date. Now by today's standards that's not too rakish, is it?

Oh, well, why comment further? In our free society, you can say what you want, but these guys are overdoing it. I'm getting too much exposure. The newspaper critics are now complaining that they're getting sick and tired—and bored—with all this coverage of me. You know what? So am I.

Incidents like these get my back up. Fortunately, they're flukes. The press has its share of bad apples, but in my experience most reporters do their homework and act responsibly. And I'm grateful as hell for them. In any event, since I write a syndicated column now, I guess I'm also a member of the press, so I shouldn't be too hard on the profession. (Mario Puzo, however, once warned me: "Hey, kid, first you write a book. Now you write a column. Why don't you stay out of my racket and I'll stay out of yours!")

More than most businessmen, I know the value of good communication and the importance of the media. There were times in our dark days at Chrysler when we had no weapons left in our arsenal except our ability to communicate. We had to keep our employees together and our dealers on board. We had to persuade the government to help us. We had to get more than four hundred banks to lend us money—when we were losing $6 million a day. We needed special deals from our suppliers. Above all, we had to convince America that we were going to make it: We had to get people to buy our cars when they didn't even know if we'd be around to service them.

That was our most important communication job, and we couldn't do it face to face. We had to go on the tube and we had to use print. I truly don't think we would have survived if our crisis had come along before the media age—we'd have died at the starting post.

Today, when reporters read my dossier at *Time* and *Newsweek*, they ask me: "Who in hell did you bribe there?"

"You guys have it all wrong," I tell them. "My method hasn't cost me a penny. You put up your money for thirty years and you bank it. You're truthful. You're available. You background reporters when they're in a jam. And guess what? You come out okay."

"Well, maybe so," they say, "but how do you come out eighty percent okay?"

"It's simple," I tell 'em. "My life's been only eighty percent okay!"

One thing I know for sure, you can't legislate the press. You can lay down some rules for Wall Street and you can get elected officials to clean up their act, but you can't dictate what appears in our newspapers and magazines. Thank heavens. Maybe that's why it's the very first amendment that protects the press—not number two or three. It may be the most important amendment in the whole Constitution.

Just imagine what a muzzled press would be like. There would be no muckracking whatsoever. Scandals such as Teapot Dome and Watergate would have remained among the country's best-kept secrets. You'd never read criticism of anything—not even movies or TV shows.

Luckily, our courts have saved us from a sugar-coated news diet by standing solidly behind the press. Around the turn of the century, one of the worst singing acts ever to appear in public was the Cherry Sisters. They were so bad that people bought onions and rutabagas from sidewalk vegetable vendors to pelt them. Critics of the day practically competed to see who could give the sisters the worst reviews. When the *Des Moines Register* really panned the girls in 1901, the Cherry Sisters, bad as they were, decided that they were going to put a stop to this stuff. They sued the paper for libel.

In one of the most important cases establishing the right of the press to "fair comment," the Iowa Supreme Court came out strongly in favor of the paper. If it hadn't, newspapers would have had to praise the dreadful sisters—and on top of that, the vegetable vendors would have lost some valuable business.

Unlike America, many other countries have always thought it was perfectly okay to control the press. That's certainly the philosophy of South Korea. Last year, word leaked out about some "requests for cooperation" memos that are issued daily by the South Korean Ministry of Culture and Information to that country's newspapers. Here are a few excerpts:

> Do not report on . . . the fact that a total of 15,000 farmers staged demonstrations on thirty-two occasions this year, the largest resistance movement of farmers since the revolt of the Tonghak Party.
>
> Treat the demonstrations of Seoul National University students critically.
>
> Do not report the UPI story on the possibility of South Korea's economy worsening in 1986.

Do not put on the front page reports of the Philippines election but carry them on foreign news pages.

Regarding the kidnapping of a South Korean diplomat in Lebanon, do not speculate on what terrorist faction did this. Report the incident in such a way as to create the impression that it was done by North Korea.

How would you like that kind of a press?

Americans are protected from such abuses. When we read something in the newspaper that seems wrongheaded to us, we get indignant. How can the reporter get away with that? we ask ourselves. Well, don't assume that the First Amendment guarantees us *accurate* news. It doesn't say a word about accuracy. It doesn't say a word about fairness. And it doesn't say a word about truthfulness. What it says is "free."

But one of the greatest things about a free press is that just as Americans never agree on anything, neither do the media. If you pick up three different newspapers, you'll find three totally different slants on the same story.

So I've got a simple solution to the shortcomings of the press: Don't believe what you read in any one newspaper. Read five. Don't swear by the *New Yorker* cartoons. Look at some other magazine's. Don't always listen to Dan Rather. Check out Brokaw and Jennings too. Now and then, when you don't like something you read, dash off a letter to the editor. You are the ultimate deterrent to a stupid press.

Whenever I give a commencement speech I always remind the students: "Think for yourselves." It sounds trite, and it is, but it's also the best advice they'll ever get from me or anybody else.

Recently I ran across an astounding finding. A survey by the Television Information Office found that half of the American public lists television as its only source for news. That's more people than ever before. How are you going to be able to think for yourself if all you've got to work with is twenty minutes of fires and Jessica Hahn interviews?

Personally, I'm an avid reader of newspapers. When I visit

another city, I make it a habit to buy a copy of the local paper. I always look at the car ads, of course, but I read most of the news too. I read *The New York Times* and both Detroit papers every day. And I mean, almost every word. I look at the pictures and the ads, as well as the articles. I skim *The Wall Street Journal* fairly regularly, and I pick up the *Los Angeles Times* when I'm on the Coast.

In addition, I usually watch Tom Brokaw on TV, though I look at the other channels as well. I catch Bill Bonds on the eleven o'clock news in Detroit. As regularly as I can, I read *Time* and *Newsweek*, and a little less regularly I read *Business Week*, *Fortune*, and *Forbes*. And I skim the *Automotive News*.

Diversity can make all the difference in how informed you are. In Detroit, we're pretty lucky. We have two newspapers, the *Free Press* and the *News*. The *Free Press* tends to be liberal-to-radical on some issues, while the *News* is so conservative you'd swear some days the dateline should read more like October 15, 1888! The two papers are as different as night and day, and yet at times they can both be so wrong it's almost ridiculous. Still, that's why we're lucky. I just take an average of the two (with a grain of salt), and the truth wins out somewhere in the middle.

Once you've looked at all the pros and cons, you have to come down in favor of keeping the press free. For all the beatings Thomas Jefferson took, when push came to shove he always stuck up for reporters. "The basis of our government being the opinion of the people," he wrote, "the very first object should be to keep that right; and were it left to me to decide whether we should have a government without newspapers, or newspapers without a government, I should not hesitate a moment to prefer the latter."

Spiro Agnew, no fan of the press, used to complain that journalists are "elected by no one." But the truth is they're elected by their readers and viewers every single day. If I don't like a particular newspaper I can simply stop buying it. As powerful as the TV networks have gotten, I'm even more powerful because I've got the best invention in the history of the galaxy—remote control. Sometimes there's so much tripe on one of the news shows that I just click off the anchorman in midsentence. Then if some guy is giving me pure baloney, he's the loser, because before he can finish his baloney, I'm already watching his archrival on the next channel. Over the years I used to use my remote control to tune out the commercials. Now I find myself blacking out the news in favor of the commercials!

Remember, in the end, if you don't like what you read in the newspaper, you can always crumple up the paper and wrap the fish with it. And if you don't like what's on the tube, you can just hit the button. It's all in your hands—literally.

IX
WASHOUT IN WASHINGTON

Will Rogers said he never met a man he didn't like. Well, if he lived in Washington today, there are a whole bunch of people I'd like to introduce him to.

I've been visiting Washington for years, and every time I leave I try to remind myself not to hurry back. At least not without a hefty supply of double-strength aspirin. It's one frustrating town. I can understand why, when Washington was founded, foreign diplomats had to be paid extra before they would agree to serve there; they insisted on combat pay.

It's tough down there—really tough. Every place you turn, somebody is arguing about something. I know checks and balances were put into our system so that there wouldn't be too much power concentrated in one place, but now there seems to be so little power concentrated anywhere that it's impossible to get anything done. Those guys can argue for months over whether to plant a new shrub outside the White House.

For all the meetings held by bureaucrats, for all the legislation from Congress, for all the proclamations by the President, the nation's problems don't seem to disappear.

Democracies are built on debate and dialogue. That's what makes them work: lots of give-and-take. But now it seems all we do is finger-point: "The Democrats are big spenders"; "The Republicans favor the rich"; "The defense budget is bloated"; "Don't you dare touch Social Security"; "Just try to raise taxes (make my day!)"; and on and on. Meanwhile, Rome is really burning.

Do we need some huge crisis to make us close ranks? Whatever happened to the spirit of compromise? What happened to America's sense of fair play? Now it seems all we do is choose sides. You're either a flaming liberal or a fuddy-duddy conservative. You're either a free trader or a protectionist. You're either a hawk or a dove. Hell, aren't there any shades in between?

I would like to see the administration take off its rose-colored glasses, and I'd like to see congressmen put on their "America" hats instead of their "state" or "district" hats. They might like the fit.

But that requires a little direction from the top. The tone in Washington, just like the tone in a corporation, is set by the guy who's at the wheel. And in this case, the helmsman, the boss, the CEO, the President, the Commander in Chief all rolled into one has been Ronald Reagan.

One of my first memories of Reagan as President was a Sunday evening in late 1982. I was invited, along with a handful of others, to an informal dinner at the White House. It was a pretty motley, though mostly conservative, group. Among the guests were George Will, George Bush, Irving Kristol, George Shultz, and a couple of others. Don Regan, then Secretary of Treasury, was left out. I would later pay for that indiscretion.

The invitation told us to come in sport clothes, just show up and be ourselves. The idea was that we'd bat around some of the problems of the day, and that way the President would get a little backgrounding without having it first filtered through his staff.

The first half of the evening we talked trade and economic affairs, because Prime Minister Nakasone of Japan was coming to town the following day. Shultz and I did most of the bantering on that subject, but in a good-natured way. I like George Shultz. He's smart, tough, and likable, and probably the best of the lot in this administration.

As you would expect, he espoused free trade forever, and I said we should manage it in our own self-interest. We did a little union-bashing, too, coming down hard on the car and steelworkers.

If the intention was to have a lively discussion, then the President sure got what he wanted. I even got a couple of licks in on the deindustrialization of America and our need for some kind of industrial policy.

Throughout the evening, Reagan hardly made a peep. It was pretty amazing to see this great apostle of laissez-faire free enterprise just sit there as I mouthed off about a little central planning. He never cleared his throat and said, "Hold it one minute—you don't know what you're talking about. You want more government and I want less. I'm right and you're wrong." He didn't try out his principles on me even once.

But you know what I really remember about that evening? Everything else is a little blurry, but this is crystal-clear. At the end of the dinner, Reagan had dismissed all the help. The President himself was walking around the table pouring us French wine. I can still hear him whispering to me as he was filling my glass: "Don't tell anybody, but I found a couple of good French bottles lying around. My friends from California would probably go nuts if they knew I served French wine to you guys."

This, I decided right then and there, was vintage Reagan. A warm, amiable, fun-loving man who didn't want to ruin a nice Sunday dinner by arguing about unpleasant issues. There's no way on earth you could dislike him. Talking politics, philosophy, or economics in his presence almost seemed uncouth. But if you really want to capture the character of the man, there's another story that I rarely share with anybody.

It was May 7, 1983, and my wife Mary was in intensive care in a coma. She would die eight days later. I came home from visiting her at the hospital, feeling very sad, and I found a message to call the President. The number was some funny exchange I didn't recognize. I thought it had to be a hoax, but I dialed the number anyway and some guy answered and said, "Oh, yeah, this is so-and-so of the Signal Corps. The President's in Santa Barbara and we're patching you through to him."

Then another guy said, "Hold on a minute. He's out chopping wood."

This is nuts, I thought. *It's gotta be a gag.*

I waited, I'd say, a full two to three minutes, and then, sure enough, on came the President. "Lee," he said, and he was breathing heavily, because he really had been out chopping wood, "I called you because I wanted to tell you how happy I was about you paying off the first four-hundred-million-dollar loan. I think that's fantastic. But that doesn't seem significant to me today. I was told how sick your wife is and I really called because Nancy and I want to tell you that we're going to pray for her. We know you're having a tough time of it and we're thinking of you."

And that's all he wanted to say. I developed a fondness for the man as a warm and wonderful human being, and I'll always feel that way.

I wish I could say the same about him as President. After that dinner at the White House, I saw Reagan on a number of occasions. And what became very clear to me was that, nice as he was, he was totally incapable of focusing in on any issue. Anytime you talked to him about something, he would drift off into an anecdote. And once he gave the punch line, the show was over and you were out the door.

Shortly before Liberty Weekend and because of the Hodel-Regan flap over the Statue of Liberty, I went to Washington. I thought it would be a good idea to drop in to see the President and ease the tension a little, because we'd be seeing each other for three full days and nights during the July celebration. The problem was getting into the Oval Office with Don Regan controlling the door. Well, God bless George Bush. He worked out a little deal with the President. Bush said I would be dropping by to see him in a couple of days and made sure Don Regan was present when he said it. The President winked and asked Bush to have me pop in. Without this charade, Bush said, there was no way he could have gotten me in.

When I did arrive, a nervous Regan joined us and began

spouting off about the President's appointment that morning of a good Italian boy by the name of Antonin Scalia to the Supreme Court, as if to say "Some of our best friends are Italian."

What I remember most is that nobody asked me to sit down. The President stood up, shook my hand vigorously, and thanked me for raising $300 million. He said he was looking forward to July. I mumbled something about the private sector coming through and how this volunteerism effort had been such a huge success.

That was music to his ears. He gave me a big grin and said, "That's beautiful, Lee. That's what has made America great."

I told him we were a little concerned about security because of the recent terrorism, and he told Regan and Bush to look into it. That was the end of that. I never heard a word.

Then I mentioned to him that we did have a few problems. "David Wolper, the great Hollywood impresario, is going to put on an unforgettable closing show that will be bigger than the 1984 Olympics," I said. "But he's getting a little carried away, Mr. President. The bills are beyond belief now. I'm afraid of going way over budget." (That may have been the first time anyone in that office ever worried about going over budget!)

And he asked, "Really? What's Wolper doing?"

I told him that, in typical dramatic Wolper fashion, he had ordered two waterfalls for the closing ceremony at a cost of a million dollars a waterfall. It seemed to me that we could somehow manage with one waterfall—that was still plenty of water—and save a million bucks.

The mention of a waterfall immediately reminded the President of an anecdote. Before I could get in another word, he began to regale me with a story about a director he once worked for. It seems that they were planning the close of a movie—a big song-and-dance extravaganza that featured a huge waterfall. Reagan thought it was pretty impressive having all that water rushing down, but the director wasn't satisfied.

"I don't think that's so hot," the director said. "If you want impact, you should have a guy go down over the waterfall in a barrel."

Then one of the assistant directors said, "Hell, you think that's dramatic, why not have a guy in a barrel go *up* the waterfall. Then we'll really have a smash."

Everybody in the Oval Office laughed uproariously. Still chuckling, they showed me to the door and said goodbye.

As usual, as soon as I got to the White House lawn, I was mobbed by reporters wanting to know what I'd seen the President about. I told them I'd talked over the plans for the Fourth of July. Then when I got into my car, my guys eagerly asked me how it went. "Terrific," I said. "All we talked about was a waterfall and a guy in a barrel going up the wrong way."

A lot has also been said about the President's lack of depth and span of interest. Some people have maintained he's incapable of conducting a meeting without having all the pertinent facts written down for him on little index cards. Well, his people do program him a bit too tightly at times. I've had my own encounter with the cue cards.

I went down to see President Reagan a couple of times on trade matters. On those occasions, at least he asked me to sit down. For me, there was a lot at stake. In one meeting I had come to talk about the Japanese beating our brains in, yet he began by pulling out a three-by-five card and reciting statistics about the GNP, auto sales, and various other economic indicators. I thought to myself: *What the hell's he reading to me for? I'm the one who's supposed to be nervous. I should have the cue cards.*

I don't know how, but word of that incident spread pretty quickly, because not long afterward Walter Cronkite came to interview me for a show he was doing on technology. During a break in the filming, Walter said, "Listen, let me ask you something. I heard that the last time you were in the Oval Office, the President opened a drawer and there were all these little index cards in the bottom of the drawer. As he talked to you he would open the drawer, quote a couple of numbers, and then slam the drawer closed."

That killed me. I told Walter that there had been one time when Reagan had a three-by-five card and, yes, it had jolted me, because he can certainly talk to you without cue cards. He just doesn't seem eager to learn anything new. George Will once remarked about him: "He's the least curious man I've ever met in my life."

Tip O'Neill saw it a little differently. He said, "He's the least-informed man I know. He does not like to do homework. He does not like to study. He does not like to read newspapers."

I guess there's nothing wrong with that—if you're a welder or a chef. But, Christ, I think most people might feel a little better if the guy in the White House used the reading lamp at least once in a while.

One day when I was talking about Reagan with my people at Chrysler, I said, "You know, in life there are some people you meet who you'd just like to pal around with. They're fun, they put you in a good frame of mind, and they make you feel terrific. Well, that's Ronald Reagan. He's not Dr. Strangelove—he's Dr. Feelgood. I want him to be my pal. But not my President."

I continued to see the President a couple of times a year until the day Don Regan appointed himself chief of staff. And then they locked the doors at the White House and pulled in the welcome mat.

Regan and I were never destined to become bosom buddies. In fact, almost from the moment we first laid eyes on each other we were like oil and water. As Secretary of the Treasury he also became chairman of the Chrysler Loan Guarantee Board, so I had to deal with him on a regular basis. And from day one, he made me understand that he wasn't going to lift a finger to undo anything that Jimmy Carter had done. In his first week in office he told the press that he was going to view with suspicion whatever Chrysler did. After all, hadn't we violated all Wall Street stood for by going to the government in the first place? He was actually annoyed that we were starting to get well.

Regan was Mister Tough Guy. I think I'm pretty good at first impressions—a big part of my job is to size up people quickly—and I sensed from the very start that this was a tough marine colonel who never listened well. He wanted everybody to know that he was tough, or he couldn't have been a marine in the Pacific. He wanted everybody to know that he was as shrewd as they come, or else he couldn't have risen from mailboy to become head of Merrill Lynch.

To this day I have no idea how Reagan ever appointed Regan in the first place. He might have come in and told Reagan an Irish story and the President said, "Hey, I'm Irish, too. Let me tell you a dilly about Pat and Mike." Knowing Reagan, I have a feeling it was as easy as that.

Regan was a smart cookie, though. He saw Ronald Reagan for what he was. He realized that the President was so laid back in management style that if Regan could become chief of staff, he literally could run the presidency of the United States.

So one day he got Jim Baker, then the chief of staff, into the back room and said, "Hey, Jim, how would you like to switch jobs with me?"

And Baker said, "Gee, I've been wanting a Cabinet post for a long time. This triumvirate with Deaver and Meese ain't too rewarding."

And so they pulled that now-famous switcheroo. Which brings me back to my moral: Never let people switch jobs at a high level just for the hell of it. It's a fundamental rule of good management. In the old days, Chrysler used to do that all the time with its executives, and that was one of the big reasons it got into such deep trouble. But if two guys on their own ever ask you to switch jobs, beware. That's an even more fundamental rule of good management.

Once Regan became chief of staff, his basic message to me was: "I'm running this administration now and you're not welcome here under any circumstances." Every chance he got, he'd try to get in a needle. After the Statue of Liberty firing, the press asked Regan about me, and he made the crack: "I don't dislike Lee Iacocca. I just don't like his products."

Come to think of it, I didn't think much of the product he was turning out either.

To give you an idea of Regan's pettiness, early in 1987 there were eighty-two cars in the White House car pool. Eighty-one were Chrysler Fifth Avenues. Don Regan's car was an Oldsmobile. Later, I found out John Poindexter also had an Olds while he was there, which gave me a feeling the two of them had to be close friends. If not on Irangate, at least they saw eye to eye on the cars they drove.

If only because I was so biased, the day Regan became chief of staff I told everybody I ran into: "That guy is going to get this country into one hell of a mess."

I also tried to tell some of the people close to the President. Holmes Tuttle, an old friend and Lincoln-Mercury dealer from Beverly Hills, has long been one of Reagan's closest associates. For years Tuttle told me he was going to make Ronnie governor of California, and I used to laugh. After two terms, I stopped laughing. Later, he said he was going to run Ronnie for President. After two terms, I am definitely not laughing.

Well, in November of 1986, Tuttle called me and said, "I'm eighty-one, Lee, and Ron and I have been friends all our lives. Before I die I want to make sure that his presidency goes well and we build a great library to a great American. That will be his legacy and then I can die happy."

Which was his way of hitting me up for a million bucks for the library.

When I heard the figure, I just about choked. We did give $50,000.

But the reason I remember the conversation so well is that after I told Holmes I'd get back to him on the million, I said, "While I have you, I really want you to know that I think Don Regan is going to get Reagan in deep trouble. If you really care about his legacy, this guy has got to go."

Well, ten days later, the Iran thing hit. The next day I put in a call to Tuttle late at night. It turned out he had just left California but would call me when he arrived in Florida.

We ended up missing each other, but I wasn't calling to say "I told you so." I was calling to tell him: "Remember the night of the library talk? Well, Regan's now got your man into real trouble, and so you should urge the President to go on television and tell the truth. Say: 'I made a mistake,' or, 'Since I'm the boss, somebody made a mistake for me but the buck stops here. What happened was awful, but it'll never happen again.' Then wait seven days, a nice mourning period, and fire all of them—Regan, Poindexter, everybody. And don't explain why."

Unfortunately, I never had that conversation because I never caught up with Holmes. But I remain convinced that Reagan's staff let him down—and in the worst way you can think of.

The next day, I couldn't believe my ears—or my eyes. Reagan went on TV and held that famous press conference in which he said he didn't do anything wrong and old Ollie North was a hero. And Don Regan is such a nice fellow, he'll keep him around till he dies.

Why didn't he just take out an ad and say: "Hold it. We screwed up. We're sorry. And it'll never happen again"?

Reagan's problem was that he worried about his image. He worried about the Republican party. He worried about his Cabinet guys, his California mafia. All the wrong stuff. He didn't worry about his customers—the American people.

I guess the President was lucky to have somebody like Nancy Reagan, because if not for her, Don Regan would probably still be running the White House. I don't want to use the old cliché that she wears the pants in the family, but close to it.

Many people have compared her to Edith Wilson. When Woodrow Wilson had his stroke, his wife basically ran the government. I don't think Nancy's running the government, but I also don't think that when the two of them go upstairs at night, she says to the President: "Don't talk business with me. Keep your problems at the office to yourself." I think she's involved. And if she was the only person in Washington with the sense to get rid of Don Regan, I applaud her.

Despite my dislike for the man, I must say that even I felt a little compassion for Regan after he got canned. I've never seen a guy at that level of government leave without even his wife saying a nice word about him. Not one press guy, not one congressional man, not one of his aides, not one of his peers had a single kind word. If there's one thing in my life I remember, it's how I got canned. When I saw the footage on TV of him on his last day, wearing old clothes and loading his stuff into his car—even though it was that damned Oldsmobile—I had to feel sorry for the way we dispose of people.

During the whole Iran mess, Reagan demonstrated beyond any doubt that he's missing one basic gene in his body that every chief executive must have. When you delegate to someone and he really messes up, you usually have to fire him. If a guy lets you down badly, he's going to wreck you. And then you're not chief executive anymore. I'm not saying it's fun firing someone. Anybody who finds it pleasant has got to be a little demented. But sometimes you have to do it—for everyone's good.

Reagan has never been able to fire anybody around him. He didn't have too much trouble canning eleven thousand air-traffic controllers and jeopardizing the safety of the skies, but with his aides he's an absolute loyalist. Show me a loyalist who gets emotionally involved with his people, and I'll show you a lousy chief executive. I'm convinced Reagan never would have fired Don Regan.

A lot of Presidents over the years have had a hard time getting rid of the eyesores in their administrations. When Sherman Adams, Dwight Eisenhower's chief of staff, admitted that he took a vicuña coat and a few blankets from a textile manufacturer, Ike kept him around until Republican leaders warned him that fund-raising would go to hell unless Adams was kicked out. But no one quite matches Reagan in the blind loyalty department.

I'll never forget when *Fortune* magazine interviewed Reagan a couple of years ago. Since *Fortune* is a business magazine, it questioned him about his management tips. Then it published this piece with its usual hype as to how fabulous a manager he was. In the interview Reagan said what every manager says, including me: I always delegate. I'm running a big operation, but I'm pretty relaxed about it because all you have to do is delegate.

Fortune seemed to think that was about the most profound insight ever uttered. But it forgot to ask the President one additional question: "Delegate to whom?"

It's easy to delegate, but when you give a job to somebody, you better make sure he knows what he's doing. He has to understand the objectives. You've also got to follow him, almost as if you were a detective putting on a tail. You've got to remind him periodically of the assignment. And you've got to find out exactly what he's doing about it.

Reagan did none of those things. He delegated to a bunch of

stooges and then went off and rode horses. He assumed that those people were doing a good job without ever bothering to find out.

There are three floors in the White House, and Reagan turned over all three to somebody else. Mike Deaver was the second-floor man. Ollie North was working the basement. And the main floor was run by Don Regan.

But what would you expect? If the boss is laid back and unquestioning, then anything goes. How else could we give birth to three guys like those who infiltrated the center of power itself—the actual White House?

Reagan delegated to these guys and then turned off the lights and went to bed. While he slept, they were busy circumventing the Congress and peddling influence. At Chrysler, I delegate, and often I wonder if I really know what's going on at some of our factories—or even in the five floors of our home office. But in my own home I don't worry at all. I would never lose control of it. Reagan did.

And when he finally did discover that his buddies were screwing up royally, he never fired them. Well, then he failed the acid test for a CEO.

A CEO has two jobs. The first one is to be able to spot good people to help him run the organization. The second job is to make a profit. In order to make a profit, the CEO has to have a track to run on, and so he establishes budgets and then he lives with them.

It's interesting that by law all the governors in the United States must balance a budget every year—except Vermont. It doesn't have a law, but it hadn't had a deficit in two hundred years. Yet, because the law didn't exist there, during the recession of 1980–1981, Vermont became the first state ever to run a deficit.

Of our first fifty presidential elections, thirty-five have had a governor or ex-governor running. And seventeen of the governors won. But the odd thing is that as soon as they got to Washington, the very principle that had been their lifeblood—balance your budget—they forgot.

Reagan seemed to forget while he was still on the plane to Washington. And nothing ever jogged his memory.

* * *

The country ought to be run like a company, but unfortunately it is not. In a company, we're problem-solvers; the government is a problem-prolonger.

At a dinner not long ago with Howard Baker, Don Regan's successor, we got into a discussion of what's wrong with Washington.

He said, "I don't know, Lee. It's not like the old days. It's almost impossible to govern. With the partisanship in the House and the Senate and the Executive branch, everyone is so uptight about keeping the process pure that nobody cares about the results."

The point he was making was that back then, when an issue got into gridlock everyone went into the back rooms, smoke-filled or not, and cut a deal. That sounds bad, but it got rid of the gridlock and it produced results. Now you can't function that way. If anything is kept secret, even for a day, then somebody will get on your case and start an investigation—or even a witch hunt.

There's always been tremendous contentiousness between the parties. But it used to be that when you got to insoluble issues you said, "Hold it. The public out there wants some answers. Let's make a deal and move on." Now it's talk and talk and still more talk.

Before I left Baker, he asked me if there was anything the government could do quickly to help the car industry. I was getting pretty depressed about what he had been saying about the problem-solving process in Washington, but when he asked if I needed anything, I perked right up and said, "You bet I do."

I brought up a problem that, for a change, Ford, General Motors, and Chrysler were united on. The three of us had a solid case that the Japanese were dumping pickup trucks in this country by selling hundreds of thousands of them below cost. They were paying a 25 percent duty, they were giving away thousand-dollar rebates, and the yen was at 140 to the dollar. What the hell were they building the trucks for—$49.95?

Through normal channels in Washington, it would take a minimum of four to five years for the bureaucrats to come up with their findings to prove injury in a dumping case. By the time that happened, I'd be retired. And the Japanese would still be merrily shipping boatloads of trucks into our market every day.

Considering that our domestic truck industry is a big employer, a big taxpayer, and a big voter, it seemed to me that the President might be more than casually interested in the problem.

So I told Baker that all he had to do was to get the President to pick up the phone and say, "As chief executive I want to know who in my administration is handling truck dumping, and I want the three of them or eight of them in my office tomorrow morning at eight-thirty."

Then he could handle the matter just the way we do in business: "Boys, here's the problem. The Japanese are dumping. Is that true or false? You've got one day to come back and tell me if the facts are right. If they are, I want it stopped immediately."

Well, I gave Baker my spiel and he promised he would pass it along. Nothing ever happened. And that's just one example. I've never been more reminded of the old saying: "Politicians and roosters crow about what they intend to do. The roosters deliver what is promised."

It's startling to see the difference between the politicians in our state capitals and the guys in our country's capital.

The states have had programs for a long time designed to attract new industries and support old ones. (In fact, four governors covered me for $207 million in loans that helped Chrysler survive a few years ago, and that kind of thing you don't forget.)

Today, I see fifty states fighting for jobs, but I don't see that same fighting spirit in Washington. The fifty subsidiaries are waging a decent battle, but they aren't getting much help from the home office.

Every state seems to have a foreign trade policy, with missions going all over the world to increase exports and attract new jobs. But I've looked for a long time now, and I can't find a policy in Washington that promotes the same determination to compete.

Maybe it's simply because governors are closer to the problem. They see firsthand the human tragedies that come from an America that seems to be losing its competitive edge. They live, up close, with the realities of a ghetto or a town where the factory just closed. And they have to care, because over and above the human suffering, their tax base changes every time one shuts down.

I couldn't help but notice over the years that there just aren't many ideologues in the statehouses. They've got to get the potholes filled. There are plenty of practicing ideologues in Washington, but not in Lansing, or Columbus, or Springfield, or the other state capitals. I don't think many of our governors tuck themselves in at night reading Adam Smith. Eighteenth-century economic theories don't help them explain to their constituents why the local steel mill is going belly-up.

Some of them really got pragmatic a couple of years ago when the GM Saturn plant was up for bids. They fought like tigers. They tried to outgun one another with free land, free training programs, free roads, low-cost energy, tax abatements—anything it took.

Then I thought to myself: *This is really strange. Nobody in Washington is fighting for American jobs, but we've got the states fighting among themselves for jobs.*

I know I'm beginning to sound like a President-basher, but look at what Reagan's left us. He's left us with a polarized society. He's left us with too much food over here, people starving on the street over there. He's left us with a young guy from Harvard making his second million dollars on Wall Street while another guy who's the same age gets laid off from a steel plant. He's left us with the eighteen biggest mergers in the history of the country, and the jury's still out on what good they did.

We sure don't want to become a socialist society. But we also don't want to end up without a middle class. That's what Brazil and Mexico did. There's squalor sitting right next to magnificent estates. What kind of countries are those? And what kind of governments allow that?

I've tried to come up with things to say in Reagan's favor—and I've really had to rack my brain. I have to assume that when he came into office, the defense budget needed some repair. He repaired it with a ton of money. It was like putting two new roofs on a house so it won't leak again. In foreign affairs, I don't know that he's moved the needle much. It seems as if we've got as much confusion now as we've ever had. You may say, "Well, you've got to give him credit. We're not at war." Okay, I'll give him that one, but he's even flirting with blowing that.

From the minute he was first elected, Reagan always talked about his vision of what he wanted America to be: strong and prosperous and free. Well, I happen to share that vision. And I'm sure everyone else does too. There's only one difference. The President thinks we're almost there. And I see that dream fading farther and farther into the distance.

There's no denying that Reagan has been extraordinarily popular. The popularity polls always ranked him ahead of the Pope. Every year he's been in office, he's been *numero uno*, the most admired man in the world. But who said a President has to be popular? Truman's rating when he left office was lower than Nixon's—and we all remember how *he* left. Yet today Truman's among the most esteemed.

One of the reasons everyone liked Reagan was because he wanted to be liked: He always went by the polls. This President didn't dare go to the bathroom—much less issue some executive order—without first checking to see what Gallup had to say. And that's no way to run anything, not even a paper route.

I always had to scratch my head when I saw the polls which showed that Reagan's personal popularity was way above the popularity of his stands on issues. How could his personal credibility be higher than the support for his positions? It doesn't make sense.

If nothing else, Reagan has certainly been true to his colors. He is just about as ideologically pure as anybody who's ever held his job. He's going to sit in heaven with the virgins, believe me.

When the historians get done picking over the legacy of Ronald Reagan, will he go down as the Great Compromiser? The Great Emancipator? (He sure won't go down as the Great Economist, not with the budget and trade deficits running wild.) No—he'll forever be the Great Communicator. And that doesn't cut it in the long run. I doubt he'll make the top twenty of American Presidents.

In the end, he's been a hands-off, laid-back leader. I don't think he's got a bad bone in his body. I don't think he would ever break a law. It's just that he didn't do his job well.

He once said, "I've been in the movies all my life. I got paid to give people two hours of feeling good during the Depression days. They came out of the cruel world and into the heated movie house. It was a nice movie. The good guy always got the girl and the good guys always won. And that was my life."

The problem is this double feature ran a bit long. All the popcorn is long gone and yet this eight-year double feature is still grinding on.

SCANDAL, SCANDAL EVERYWHERE

X
FREE TRADE OR FREE RIDE

If you're wondering what I've been doing in the four years since I last wrote to you about the trade mess, the answer is *worry*. Because things have gone right down the tubes.

In a nutshell, here's what's happened: The war with Japan is four years older now, but the Japanese have won every skirmish and every battle. All we've been doing is waving a white flag.

And that's not all. Now the rest of the world is saying, if it worked for Japan, why won't it work for us? So you've got South Korea flooding our markets. Taiwan is going like gangbusters. Brazil is banging on the door. Even Rumania is shipping cars over here. Then there's the Yugo from Yugoslavia. And the Proton just coming in from Malaysia. If you had told me four years ago that Yugoslavia and Malaysia would be shipping cars here, I'd have said you were nuts. Hell, now even Russia wants in.

The whole world wants to come to America. Of our twenty largest trading partners, we now have huge deficits with seventeen. So this has become ye old dumping ground. This is the big Turkish bazaar. Set up a stand here, sell your wares, and cart the money home.

You would think that by now we would have figured out a simple plan to compete in the world, but we haven't. We didn't have one four years ago, and we still don't. Every other country has a plan to manage its trade; we're the only dinosaurs left who maintain the myth that "free trade" really exists. We hope that somehow things will just automatically straighten themselves out, that some invisible hand will wave the magic wand of the free market. Talk about innocents abroad! When it comes to trade, we have become the patsies of all time.

And here's why I don't understand it. Not long ago our own government published a book listing only those "significant" trade barriers we face around the world—and it was 240 pages long! Some of them make me want to laugh, but most of them make me want to cry.

In South Korea, for example, a citizen can get arrested for smoking an American cigarette. We can sell American apples in Sweden, but only after the Swedish crop is gone—and by that time ours is a little mushy. Spain discriminates against our squid, Portugal against our soybeans, and Taiwan has an outright ban on "edible offal," whatever the hell that is!

We aren't as pure as we proclaim, either. We have by far the most open market in the world but we, too, have restrictions on dozens of imports, from sugar to motorcycles. We even have a ban on avocados, but only if they have pits in them. (As far as I know, every single one of them still comes with a pit.)

But the heavyweight champion of the world when it comes to regulating trade is still Japan. Japan's list of restrictions is dazzling, and it covers everything from walnuts to frozen herring. The Japanese game plan is simple: take in American raw materials but not finished goods, because that would cost them jobs. So they import American lumber, but not plywood; animal hides, but not leather goods; potatoes, but not French fries; tomatoes, but not tomato purée.

Since I'm of Italian extraction and pasta and pizza have always been staples, the tomato battle caught my eye. But now the Japanese are cracking down on ravioli. We can send them all the ravioli we want as long as they're filled with cheese. But if they're filled with meat, there's a quota. That's because the Japanese restrict U.S. beef, and I guess they're scared we'll smuggle it in in all those little raviolis. These guys just don't miss a trick.

You know, I hate to pick on the Japanese—but they give me such great material to work with, I just can't help myself!

Sometime in 1993 the new Kansai International Airport will open near Osaka. It will cost as much as $8 billion, which is a mouthwatering number if you happen to be in the construction business. So when the airport was announced, a number of American construction companies went after a piece of that $8 billion, only to be told that, sorry, they couldn't bid.

The reason given: "Only a Japanese construction company understands Japanese soil conditions."

When I heard that, I said to myself: *These guys must be kidding.* We've got a sense of humor, but this is taking things a little too far. Because at exactly that time Japanese construction companies were doing almost $2 billion in business in the United States. They were building everything from auto plants to hydroelectric facilities, while American construction work in Japan was limited to an occasional hamburger stand.

A week or so later, our ski guys were told that they can't sell skis in Japan. Reason: "Our snow conditions are so different from American snow conditions that American manufacturers could not meet our specifications on safety."

One week it's soil. The following week it's snow. Next week they'll find something in the air or the water. One thing's for sure: They're making fools of us.

Foreign cosmetics companies must test their products in Japan before actually selling there because the government insists that Japanese skin is different. The government also maintains that Japanese stomachs are so petite that they have enough room for only the local tangerine. That's their imaginative excuse for why U.S. orange imports are limited.

When an American retailer tried to export first-quality towels to Japan, the infamous customs inspectors turned them down. It seems the towels were so fluffy that when you twisted the corner to dry the ear, it was too bulky for the delicate ear canal of the Japanese.

While the Japanese keep slamming doors in our faces, they keep barging in on our markets. Everything is fair game. The Iwo Jima Memorial, commemorating one of the great battles against Japan in World War II, is now being pictured on postcards sold in the Senate gift shop. You guessed it—the postcards are stamped MADE IN JAPAN on the back.

And the symbol of U.S. citizenship, the passport, is now "made in Japan"—or at least "made by Japan." The new American passports are being put together by a Japanese machine and assembled by a team of Japanese sent to Washington by the manufacturer, Uno Seisakusho Co. Ltd.

I do have to admit that personally I have had some good news on the trade front. I'm happy to report that Chrysler's sales in Japan rose 272 percent in the first four months of 1987. That meant our sales shot up to forty-nine cars from eighteen. Who says we're not trying?

And Japan's Ministry of International Trade and Industry has come up with a novel way to reduce its trade surplus—some day it plans to export its old people overseas into Japanese-financed retirement villages in America. Japan figures that the building and maintenance costs, plus the money the senior set will spend in their new country, will help knock down our deficit.

One last bit of good news: The tourist business here ran a surplus with Japan of $1 billion last year. So while Japan was selling us cars, TVs, and cameras, we were selling them beanies, pennants, and cotton candy.

We've been fighting for years over baseball bats and walnuts. Meanwhile, in just over five years, we've gone from a trade surplus of $40 billion with the rest of the world to a deficit of $170 billion. Our $60 billion deficit with Japan alone is higher than any country in its right mind would have thought was a tolerable worldwide deficit— and double what it was in 1984, when I was yelling about it.

Some of the other numbers are pretty scary too. The South Koreans have gotten up to around $8 billion. They were at only half that in 1985. Today they're playing a great exporting game, just the way Japan did in the early days.

Some people say: Well, they're our allies, aren't they? Korea is split down the middle, so we've got to let South Korea prove that our system is better than the North Korean system. It's like East and West Berlin. We've got to be tolerant of friends, because we don't have that many friends in the world.

Well, just take a look at which countries we have a trade surplus with. We have a small surplus with Holland, Belgium, and Australia, but after those countries we have to rely for our exports on such lush and lucrative markets as Paraguay, Greenland, and the Falkland Islands. We do all right in Tanzania and Pakistan, too, in case you're interested. And here's the ultimate irony: We even have surpluses with Russia, Vietnam, and Libya, believe it or not.

We can have a trade relationship with our enemies, as long as we're careful not to sell them any defense-related products. Maybe that's why it's our friends who are laughing at us.

Let's face it. We're out of step with the rest of the world. And when you're out of step with the world, you don't say the rest of the world is wrong and you're right.

It reminds me of the British driving on the left-hand side of the road. They think the whole world is out of step for driving on the wrong side. And we think they're the dotty ones.

You know, rape is not known among porcupines. But if you're the only porcupine without quills in a field, watch out. When it comes to trade, we've lost our quills (and our wills too).

As ridiculous as our trade deficit has gotten, our legislators sit around and suck their thumbs. Whenever anyone suggests they do something, they worry about retaliation. In trade, the guy getting mugged is supposed to retaliate. But in Washington a legislator would rather be called a pervert than a protectionist.

You see, even though the rest of the world can get along—and they can—without our steel, and our cars, and our grain, and even our high technology now, we've got a hole card that tops anything the others can lay out on the table.

It's called the American market. That's our ace of spades. None of the major trading nations of the world today can survive without the American market. More than 40 percent of Japan's exports come to this market. We take 48 percent of Taiwan's, 40 percent of South Korea's, and 32 percent of the exports of the entire Third World.

The profit figures get even higher. The largest maker of cars and trucks in the world is no longer the United States—it's Japan. But for years now, the Japanese have gotten almost 100 percent of their total worldwide automotive profits from just one single market—ours. So what are we afraid of?

We can't afford this trade imbalance any longer. Because the price is Americans thrown out of work. The numbers are depressing: 237,000 of those who worked in steel just six years ago have lost their jobs; 182,000 in fabricated metals; 434,000 in heavy machinery, 300,000 in textiles and apparel—the list goes on and on. If you want to know what the trade deficit means, just take a look around you. I know, I know, these are "mature" industries that are dying. Our future supposedly lies in retraining those workers for all those burgeoning service industries. After all, who needs to make cars when you can do just as well washing them and gassing them?

Our friends overseas can't afford this imbalance any longer either. See, the Japanese don't sell us anything we can't do without. We don't buy their food or natural resources. We like their cameras, TVs, cars, and stereos—but we don't need them. Japan is betting that America will continue soaking up most of its excess industrial capacity, and it's becoming a bad, bad bet, because America simply can't afford to.

The two best ways I know to ruin a friendship are to go into debt to a friend or to become too dependent on him. And that's what Japan and the United States are doing to each other.

We're unfair to ourselves, and we're unfair to Japan in allowing the United States to become the world's shopping mall. It's really time to start charging admission, and the price of a ticket has to be a little reciprocity.

Last year, when the Japanese reneged on our agreement with them limiting the amount of microchips they could ship here, our government decided to play tough guy and hit them with $300 million in countervailing duties. But look at what that really amounted to—a chip shot. That $300 million represented all of three tenths of one percent of the United States' trade deficit with Japan—equivalent to a lousy seven boatloads of cars.

Not only that. After our government finally took some action, the American press made it look as if we were the bad guys. We were going to start a trade war. But no one seemed to remember that it was the Japanese who broke an agreement. The reporters' conclusion was that we'd fired the first shot in a war we can't win.

The truth is, we didn't do enough. We fired a shot with a peashooter when we needed a cannon.

I'm glad America has such a strong service economy, but there is such a thing as an economic food chain, just like the one in nature. And it starts at the bottom with people growing things and building things. Somebody in the pond has to produce something or pretty soon everybody in the pond dies. And right now, in America, the people who produce things are in deep trouble.

With the shrinkage in manufacturing, 76 percent of our labor force is now in the service sector. Now, that 76 percent may work pretty hard, but they don't do much toward solving our trade problem, of course, because services are a little tough to export. On the other hand, the service sector is a whole lot safer place to be these days, because there's virtually no foreign competition. (Banks are the big exception, and you see what's happening to them.) The fact is, there's just not much incentive to improve our productivity if we concentrate on service industries, because nobody from Japan or Korea is breathing down our necks.

We don't send our laundry to the Far East—yet. We can't call up Swissair and fly to Chicago. *The New York Times* and *The Wall Street Journal* can write editorials all day long telling people like me how to handle foreign competition—but they've never in their lives faced any foreign competition.

Eventually—mark my words—even the service industries will be lining up in Washington along with everybody else, looking for help.

Ninety percent of all the new jobs in the last sixteen years have been service jobs; but watch out, because pretty soon there won't be anybody left to service. If the industrial sector goes down the tubes, who's going to buy the hamburgers? Whose laundry is going to need washing?

I hate to admit it but I have no idea what our trade policy is. There are eight different entities in Washington creating trade policy—different trade policy! As a result, America has the full attention of its trading partners in much the same way that the cross-eyed javelin thrower has the attention of the spectators in the stands. They are a little bit apprehensive, because they aren't sure what the hell is going to happen.

My industry and its workers have been hurt more than anybody by Japan's strong trade policies and our weak ones, so I've screamed louder than anybody that we need to level the playing field. I've gotten a reputation as a "Japan basher." It's a bum rap, though. I'm not as mad at Tokyo as I am at Washington, and nobody has ever accused me of being an "America basher."

In March 1981, at the depth of the recession and with the U.S. auto industry showing record losses, the Reagan administration asked Japan to agree to a voluntary import quota of 1.68 million cars. The number increased to 2.3 million by 1985. By then, the Big Three were getting healthy again, but the Japanese cars coming in still had a $2,500 cost advantage over ours, mostly because of tax rebates by the Japanese government, and because the yen was a whopping 260 to the dollar.

With that kind of advantage, the Japanese could have flooded our market with new cars if the quotas were removed. We were already almost $4 billion in the hole each month to Japan, and $1.6 billion of that was in motor vehicles alone. So when the people in the administration started hinting that they would not ask Japan to continue the car quotas, I couldn't believe they were serious.

But sure enough, in February 1985 the President's Cabinet Council voted 12–0 to let the quotas die. In effect, they would let Japan determine how many cars it would ship to America each year. The country was already drowning in red ink, and our government was leaving the floodgates wide open.

I knew that the Japanese would exercise some restraint. They wouldn't be dumb enough to overplay their hand and risk a backlash. But they would send in all the cars they could get away with. They'd set a number that would be just one car short of the straw that breaks the camel's back. In effect, the Japanese would be unilaterally writing America's trade policy on cars, and here, all these years, I had this dumb notion that our trade policies should be written in Washington, not Tokyo.

I sent Ben Bidwell, our vice-president for sales and marketing, to Washington, just to be sure the administration understood the consequences. He told trade representative William Brock and his staff that this meant Chrysler would have no choice but to follow General Motors to the Far East for more cars and parts, and that we were going to make the trade deficit even worse. They just nodded and said, "Now, you've got it right!"

So when the quotas were taken off—on April Fool's Day—I was packing my bags for Japan and Korea. All I can say is, "The Devil made me do it." If the government was going to engineer a big jump in the trade deficit, I owed it to Chrysler to get my fair share of it. In fact, I had absolutely no choice. We looked at the numbers. We saw what unrestricted Japanese imports would do to us. And we had all the evidence we needed that the administration would do nothing to stem the tide.

Maybe what bothered me most of all was that the administration gave Japan open season on the American auto market and got absolutely nothing in return. The bilateral trade deficit was on its way to another record of $148.5 billion that year. Citrus growers, beef growers, and other farmers were all pleading with the government to break down Japan's trade barriers against their products. The cigarette companies wanted help. The high-tech people wanted help. Just about every American industry that could have competed effectively in the Japanese home market was being systematically excluded, and their lobbyists were running all over Washington pleading with our government to give them a fair shot over there.

The administration would never have a better chance to open the Japanese market. Autos were their ace in the hole, and they threw it away. Apparently these guys had never played a hand of serious poker in their lives. That's when I finally, and sadly, concluded that not only did the Reagan administration lack a sensible policy to handle the growing trade tension between America and Japan, but it also had no intention at all of formulating one. I had a gut feeling that tensions would get even worse—and I was right.

I saw those tensions in Japan. I also saw confusion. Most Japanese were happy about the President's action, of course, but they also noticed that only a few days afterward the United States Senate voted 92–0 to tell the President to retaliate against Japan for its closed-door trade practices. Talk about yin and yang: The President tells Japan: "Send us all the exports you want," and the Senate, controlled by his own party, votes unanimously to slam the door.

I spoke to a lot of wise heads in Tokyo who understood clearly that reducing our trade imbalance was essential so that Japan and the United States could maintain their close and special relationship. They seemed as worried as I was. But what were they to make of the mixed signals coming out of Washington? Just a few days before I got there, Prime Minister Nakasone went on national TV and told every Japanese to go out and buy $100 worth of imported American goods. It was hard to imagine, but it was true: An American President was sitting on his hands doing nothing and a Japanese Prime Minister was taking a big political risk by appealing to his people to "Buy American."

I've never liked the tone of most "Buy American" appeals in this country. Somehow, they sound like American business and American workers need charity, when all they really want is a chance to compete. But when I heard it coming from the head of a foreign country that had been cleaning our clocks, I really got embarrassed.

Despite Prime Minister Nakasone's heartfelt appeal, the trade imbalance kept getting worse by the month. Japan's export and protectionist lobbies are stronger than any of the special interests we have in this country. When I went there I was warned: Whatever you do, don't bring up rice. If we really had free trade, most Japanese farmers simply couldn't compete with lower-cost U.S. rice. But rice is sacred there, and so are the farmers. There is nothing sacred here, of course, about an American auto worker, or even an American rice farmer. But a Japanese rice farmer is sacred. And every Japanese politician knows that.

While I was in Japan, I spoke before the American Chamber of Commerce. It was a large, mostly American group, but there were a lot of Japanese businessmen and politicians there too. I guess they came to see the great Japan basher in action.

A funny thing happened. I made all the same points I'd made to the American politicians, with Ambassador Mike Mansfield sitting right next to me, but when I made them in Tokyo, smack in the lion's den, the roof didn't fall in. I recited the trade numbers and said America wouldn't stand for it much longer. I brought up the citrus fruit problem, the beef problem, and all the other barriers the Japanese had erected. (I did chicken out on rice, however. Even I know when to quit.)

I pointed out that part of the problem was America's, because of our huge budget deficits that overvalued the dollar, but I said that Japan couldn't use that as an excuse as long as its markets were virtually closed and its trade practices looked predatory to the average American.

Then I hit them with the zinger. I could see all the Japanese in the audience stiffen up. "If Japan is protectionist, then I think it's protecting the wrong market. It's protecting its market in Japan, when it ought to be protecting its market in America. That's the one that's really in danger, and that's where Japan stands to take its biggest loss."

I had a press conference afterward with more reporters and more cameras clicking at me than I'd ever seen before. For over an hour they bombarded me with questions. When the stories were printed and the TV clips aired, nobody called me a "basher" or anything else. A week later, Ambassador Mansfield cited my speech at length to the Japanese Diet.

I felt encouraged coming home from that trip to Japan, not because anything I said was going to change things, but because I sensed from the Japanese leaders that they understood the problem and what had to be done to solve it.

But nothing changed. Every month America's trade deficit got worse, and every month Japan accounted for a bigger chunk of it. For each trade barrier the Japanese took down, another one seemed to pop up.

I'd hoped the Japanese were going to start breaking their dependence on us in the fall of 1985, when their government went along with Secretary James Baker's effort to bring down the value of the dollar and increase the value of the yen. After all, the currency imbalance was a major reason why the U.S.-Japanese trade deficit had skyrocketed during the early eighties. But the currency fix turned out not to be the trade fix. A year later, while everybody waited for the famous J-curve to start upward, America's trade deficit with Japan was $10 billion worse. The dollar cost of the yen increased by 50 percent, and yet the prices of Japanese goods in the United States had gone up only about 10 to 15 percent.

The Japanese took the loss right out of their own hides. Honda, probably the best company in Japan today, dropped $110 million in one quarter, leaving only another $110 million for profit. Their profits got cut in half, because they swallowed most of the currency change and raised their dollar prices only slightly.

Why? Their beachhead in this market is too strong, and they're determined not to give it up.

If the Japanese are tenacious about keeping their export markets, they're just as stubborn about resisting changes at home. Japanese don't buy much, and when they do it's Japanese. They're also the world's greatest savers; they sock money away better than Jack Benny did. Their trade partners and their own government have tried to get them to loosen up, but it hasn't worked. I guess it's hard to criticize their values: Work hard, save, and if you have to buy something, give your business to somebody you know. That sounds like what my parents taught me! But it's a big problem for a country that lives on exports and is getting pressure from the rest of the world to balance its trade.

The Japanese government even went so far as to create something called the "Leisure Development Center." The whole purpose of this institution was to get people to take time off and enjoy themselves, to spend some Saturdays riding motorcycles instead of building them, maybe even just to catch up on some sleep. It was such a serious effort that the people in charge worked night and day, six days a week and sometimes seven, to persuade other people to stay home and relax. They wrote a report complaining that Japanese workers were taking only 60 percent of their vacation time. But it turned out that the people running the leisure center took only 40 percent!

In early 1985 a Japanese government commission proposed shortening the workweek to five days from six. By the end of the year, a survey found that the number of companies with a five-day week had actually gone down. The only positive development was that the typical Japanese worker was sneaking out of the office a minute and fifteen seconds earlier.

We might as well face it. Japan's traditions run too deep to be changed from the inside. We can plead and argue and threaten for years to come, but the Japanese will stay on the path they're on—a path the United States put them on after World War II—because they don't know how to get off it. It's politically impossible for a Japanese leader either to force a cutback on exports or to open the Japanese market wide enough to do any good. All the trade meetings in the world and all the good intentions of Japanese leaders won't change that.

We've also got to keep reminding ourselves that the trade problem today goes well beyond Japan.

Every time I think of Hyundai of South Korea I get shivers because they are so good. Hyundai broke the world's record for an import by selling 200,000 cars in the first twelve months after its entry into the U.S. market. That's barn-burning performance. Now Ford's bringing in a car from another Korean company. The biggest Korean company, Daiwoo, is half-owned by GM, and they're importing a car too.

The reason I know South Korea is good is because every time I mention the Koreans to our own Japanese partners, they shake their heads and frown. The Japanese are inordinately worried about them—and they have reason to be. The Koreans have the same work ethic as they do; they're where the Japanese were twenty years ago, eager to go, with their government behind them all the way.

We worry about Japan, and Japan worries about South Korea. I don't know who South Korea worries about. Probably North Korea.

My impression of Korea from my visits there is that the work ethic is even stronger than in Japan. Calisthenics at noon. Skip lunch. I went through all those Korean plants and nobody even looked up. They work like dogs. The absenteeism in a year there is what ours is in a week; nobody wants to run the risk of ending up back in the rice paddies. The enthusiasm of the work force makes the Japanese look laid back.

The average manufacturing worker in South Korea makes a little over $1.19 per hour, works more than fifty-four hours per week, and gets no overtime. Two thirds of the overall work force make about 55 cents an hour. Talk about working hard for your money.

Not only that. Lots of foreign products are bootlegged there. Even my first book was pirated—to the tune of 100,000 copies. The Koreans observe none of the international copyright laws, let alone trade constraints. So it's time to say: "Hey boys, trade is based on mutual respect and trust, and you're going to have to change your tune if you want us to keep buying your goods."

Meanwhile, Volkswagen and Ford are going to bring in a car from Brazil. And Mexico is getting into the act by shipping us a lot more cars. All roads lead to Main Street, U.S.A.

The real wild cards are the South American nations, and Brazil and Mexico are their leaders. The U.S. banks have so many loans out to them that if we tried to stop their exports dead in their tracks, our banks would go under. Which almost gives them a favored-nation status on trade, but for the wrong reason.

Now we're going to get 1 million Korean cars, 3 million Japanese cars, and 1 million others. That's 5 million cars in a 10 million American market. Oops, there goes half the market. And the jobs that go with it.

In the world you always have some enemies, some allies, and some neutral guys. As Americans we have three major adversaries—

OPEC, the Kremlin, and Japan Inc. OPEC, for instance, has its own interests at heart, not ours. Those countries don't believe in free markets and they control a basic commodity called oil. They're going to screw us at every turn. Then we have the so-called evil empire—Russia. And so the biggest part of our roughly $300 billion defense budget goes to keep Russia at bay. We're not too worried about Mexico's invading us; we're worried about Russia.

But over here is our Trojan horse, because he's a friend smiling all the time. He's not half enemy and half friend. Japan's our last ally in the Far East, I'm told by the State Department. That's why we have to treat him with kid gloves. He's the number one economic power in the world today, and he can do almost anything he wants.

As a businessman I admire him, but he also worries me a lot.

If you took the trade deficit to its ultimate, Japan would have a worldwide cartel. You know what that would be like? Let me give you an example: When a Toyota goes up in price by $1,400 because of currency changes, a Chrysler K-car looks better to the U.S. consumer by $1,400. But take a VCR. If the price goes up $200, the American buyer may want something home-grown. But the U.S. doesn't build any. Not one VCR anymore. Unless there's someone else in the world building VCRs, the Japanese, in effect, have a cartel in VCRs—and can charge whatever they like. The American consumer may say: "The hell with the Sony. It's too expensive." But what'll he buy? Panasonic? Tell me about world competition and world trade at that point. The free-trade purist, of course, will say, "No problem. We have super low-cost Korea waiting in the wings to keep Japanese prices honest."

To me, the solution to the trade problem is simple: We have to freeze our deficit with Japan in place, and tell the Japanese to take it down by a good amount, such as 20 percent per year, until we balance it. We have to give them a flat dollar limit on how much of a deficit we'll tolerate, and tell them that the limit will get 20 percent smaller next year, and 20 percent smaller the year after that.

To put this into perspective, the Japanese trade imbalance with the U.S. has soared to a monstrous $60 billion a year. In the last three years Japan's surplus with us has grown at 30 percent a year! Poor Dick Gephardt suggested we ask the Japanese to take it down a mere 10 percent a year and he was branded a radical protectionist by the press. Take my word for it, the Japanese like to work toward objectives. It's time we gave them one.

Since Japan is a sovereign nation, we've got no right to interfere and tell them how to balance their trade with us. But since we're a sovereign nation, too, we have all the right in the world to tell them to balance it. They can sell us less or buy more—we don't care. The choice is theirs. We should knock off the endless negotiations on ravioli one day, grapefruit the next. That's gotten us nowhere, and it never will. A simple limit is all we need.

That idea really constipates the free-trade purists, of course, but it meets my two criteria of a sensible trade policy: fairness and self-interest. It's fair because it's time for Japan to start paying some of the dues that go along with being a major economic power in the world. The United States has paid its dues with the Marshall Plan, and massive amounts of foreign aid, and by letting Japan and everybody else have open access to our markets without their giving us the same in return.

Japan is on its way to becoming the richest country in the world. It's already so flush that a single membership in the most prestigious Japanese golf club recently sold for $2.7 million. That would leave a dent even in Arnold Palmer's wallet.

So Japan doesn't need the free ride we've been providing anymore. We pay for Japan's defense and we provide 38 percent of its total foreign sales. We keep 3 million Japanese employed. We've done our part, and it's time for Japan to start pulling its own weight. It got rich by targeting our market and protecting its own. Now it's fair to tell them that the rules have changed.

A deficit limit also meets my self-interest criteria, because it's obviously not in America's interest to become an economic colony of Japan, and that's where we're heading. But it's not in Japan's interest, either, to continue its almost parasitic dependence on the American market.

Anything we do to right the trade imbalance, though, is going to be difficult for Japan to take. We're dealing in perceptions as well as realities. Maybe because I didn't grow up on an island, I'll be the first to admit that I don't fully understand what motivates the Japanese. But that works both ways, and it's apparent that many Japanese still don't have even a basic understanding of America or Americans.

One prominent Japanese television commentator, for example, was quoted recently as saying: "We've accepted trade friction like changes in the weather—something natural—and tried to adapt in a passive way, as you do against a storm or wind."

Between those lines I saw exactly the attitude that has caused much of the problem: The Japanese aren't serious about changing their predatory trade practices. They just intend to wait for the storm to pass.

It's discouraging enough to hear that from a leading Japanese journalist who doesn't understand America, but it's worse when the blindness extends to government leaders.

Wataru Hiraizumi has been a member of the Japanese Diet for twenty-two years and is considered an influential member of the Liberal Democratic party. He's got a simple explanation for our trade problems: "Japan is not going to change," he says. "We love to work hard and Americans don't."

Sure the Japanese work hard. Lots of people work hard, even Americans. The Japanese didn't get rich just through hard work; they got rich through open access to the American market while shutting us out of theirs. And, of course, because billions of American defense dollars go to protect them every year. In fact, not long after Hiraizumi's interview, thirty-seven American sailors died in a tragic incident in the Persian Gulf protecting oil shipping lanes. The United States gets only 5 percent of its oil from the Gulf; Japan depends on the area for 60 percent.

Here's where Hiraizumi says we're heading: "We'll continue to work hard and amass huge surpluses of money. We'll buy up your land, and you'll live there and pay rent."

That's what Hiraizumi thinks of Americans—that we're going to become a nation of tenants, leasing our own country back piece by piece from our Japanese landlords.

Mr. Hiraizumi has a lot to learn about Americans if he thinks that scenario will ever play out. But I can understand why he believes we're headed that way.

The Japanese have made so much money off of us that they're diversifying their portfolios by snapping up American real estate. You could wake up tomorrow and your landlord is no longer Joe Smith. He's Joe Akimoto. That's exactly what happened to companies named Exxon, Tiffany, and ABC. The New York headquarters of all three were recently acquired by Japanese investors.

Not only have they bought a bunch of skyscrapers in New York, they've bought a couple of abandoned casinos in Las Vegas, and they now own two-thirds of the major hotels in Hawaii. They're also in the midst of a bank-buying spree. And so they now own six of the top twelve banks in California, and six of the top ten in Chicago.

They don't even blink when you show them the price tag. The Exxon building went for $610 million, a record amount for a New York tower. In 1986 they grabbed a whopping 6 billion dollars' worth of American real estate, quadruple the total for all of 1985—and they're still shopping.

I really had to pause when Kondobo USA Inc. bought the Justice Department's Judiciary Center in downtown Washington. To its new landlords, Justice now pays $8.5 million a year in rent.

A big Japanese conglomerate took out a full-page ad in *The New York Times* last year to brag that it just had "a landmark year" and that it has "awed U.S. industry by snapping up $1.8 billion in real estate in a scant twelve months" (including ARCO Plaza in Los Angeles and the ABC building in New York, by the way). And here's the kicker line: They say to "stay tuned for future developments" because "the best is yet to come."

They really know how to rub a guy's nose in it, I'll say that for them.

Finally, the most telling and depressing thought of all. According to Hiraizumi we have nothing to worry about because: "We won't go to war. We won't destroy each other. We're *condemned* to live together." How's that for mutual trust?

I don't know how much Hiraizumi represents the thinking of other Japanese, but he's a prominent politician, and politicians (Japanese and American) rarely say anything, no matter how outrageous, unless they think it'll play with the folks back home. So I don't believe we can just toss him off as an eccentric. He and some other Japanese seem to have a certain view of themselves and the rest of the world that is pretty hard to warm up to in 1988.

We may have a long way to go in understanding the Japanese culture, where everybody has the same values and customs going back thousands of years. But it's obvious that Hiraizumi and others don't understand a pluralistic society like ours, where we'll accept anybody as our equal but nobody as our superior.

A lot of people are calling our differences with Japan a trade war. I'm getting uncomfortable with that metaphor because any time you compare anything to war, you're automatically overstating your case. Besides, it wasn't too long ago that the two countries had a real war.

It's considered impolite these days to bring up Pearl Harbor, but it's important to remember why it happened: It happened because we didn't know the Japanese, and they didn't know us.

Both countries paid a terrible price for being strangers, and—impolite or not—we can't afford to forget it.

Either of us.

XI
BUDGET BUSTERS

When I was growing up in the '40s, we used to have a lot of expressions that played with numbers. A fast car would "go like sixty." Everybody wanted to "live to be a hundred." And a "million" of anything was awesome—as in "You're one in a million," "Thanks a million," or one of my favorites, "Baby, you look like a million." (I was a big Humphrey Bogart fan.)

All I knew for sure was that a million was close to infinity. But now, forty years later, the word *billion* (that's a thousand million) has crept into my vocabulary. No, I don't go around saying "Baby, you look like a billion," but I did borrow $1.2 billion when Chrysler was dying—and paid it back. I spent a billion to bring out our minivans. And I recently signed a billion-dollar labor contract.

It's taken me forty years to comprehend a billion. And just as I'm getting the hang of it, the word *trillion* has started cropping up. (That's a thousand billion!)

Except for astronomers, nobody ever used the word *trillion*. Even our own federal government didn't comprehend it until 1981, when, after 206 years, it found itself $1 trillion in debt. But then, when the debt doubled in just four years to $2 trillion, people started to ask: "Hey, what is this?" And in three more years, when that debt reaches $3 trillion, those same people are going to get downright mean about it.

When we try to get the Japanese and others to mend their mercantile ways, they throw our own debt right in our face and say: "If you didn't have such a huge budget deficit, you wouldn't have such a huge trade deficit." And of course they're absolutely right. That's why I say that these two deficits will go down in history as the twin scandals of our time.

Let's assume, in the interest of understanding the mess we're in, that we asked our government to come clean, follow its own truth-in-lending laws, and level with us. Every year with the tax forms we would get a statement telling us where we stand on our debt. Right now it would read like this for the average family of four:

Dear Mr. and Mrs. Taxpayer:
Your share of the national debt is now $38,928.50.
In the past twelve months, your share has increased by $3,615.58.
Your share of the interest bill this year is $2,255.68.
Have a nice day.

If Americans saw the debt personalized that way, it might start a revolution. And maybe we need one.

We're handing the next generation more than anybody has ever passed on to their kids. Generations behind them were lucky if they inherited a little shack on the back forty. Our kids are getting a big, beautiful mansion on a hill.

But, just one thing before we ask them to get all choked up with gratitude. We haven't bothered to pay for it yet. I've said it before, and I wish I didn't have to keep on saying it: We're leaving them the mansion, all right, but we're leaving them the mortgage too.

Maybe we should hear from the kids on this subject. Maybe we should restrict the vote from now on to those under thirty years old, because they'll be the ones getting stuck with the bills. Along with their own debts, they're going to get the privilege of handling some of mine.

I suppose I should add, "Thank you very much," but the truth is, I'm a little ashamed. When I was graduating from college, nobody stuck me with a bill like the one our college graduates are getting—and remember, then the country had just been through a dozen years of the worst depression in history, followed by four years of the biggest war in history.

We had a big debt after all that, but nothing like the whopper we're leaving behind us now.

It's "taxation without representation" again. But this time we're taxing our own kids. And when they figure it out, they aren't going to dump tea off the dock. They're going to throw *us* in the harbor.

I'll tell you one thing: Don't try to pay off that $2 trillion-plus in cash. It would take the U.S. Mint fifty-seven years, two months, and two weeks just to print it. The interest alone on that debt is now running about $192 billion a year. That means interest payments skim 25 cents off the top of every tax dollar. Those are the kinds of rates you hear about in Las Vegas. And remember, that 25 percent doesn't pave a single road, hire a single cop, educate a single kid, or feed a single poor family. It's dead weight. How are we ever going to grow and become productive with a piano like that on our backs?

I wish I could say there's an end in sight. But if the debt keeps piling up at the same rate that it has since 1980, it's going to hit $13 trillion by the year 2000: that's fourteen times the debt in 1980.

At Chrysler, we learned the hard way what happens when you start hemorrhaging like that. Pretty soon you begin hearing death rattles. Especially when you've got no plan to turn yourself around.

The government ought to be showing the rest of us how to handle our finances so that there's always some loose change for emergencies. And yet every month the government is spending $17 billion more than it takes in. That's one way to break up a family fast.

How did all this happen so quickly? Well, a few years back somebody came up with a new ideology to handle public finance in this country. Some people called it the "supply side" approach, and others called it "trickle down."

The supply-siders said that if we cut taxes, our revenues would shoot up so fast that we could not only balance the books but could reequip the armed services to boot—and even have a lot of money to pay for all those missiles we're storing up.

"Trickle down." It sounds so easy and painless, doesn't it? But when it results in more debt in five years than the country had piled up in over two centuries, "trickle down" is going to become a water torture for somebody. And I'm afraid that somebody is our kids.

I can give you a rough idea of how "trickle down" really works. Last year the IRS seized $70.76 from a nine-year-old girl—money she had literally saved in her piggy bank—for partial payment of back taxes owed by her grandparents. After the story hit the papers, the IRS was embarrassed enough that it gave the money back. But if I were you, I wouldn't count on the IRS backing down in the future. It's going to need every penny it can get. Trickle down was supposed to cut taxes to fuel the economy, not tax kids to penalize saving.

Unfortunately, the politicians don't seem particularly worried about their spend-now, pay-later ways. Every year they play what I call the Games of August: the annual effort to fool themselves—and us—into thinking that they're actually doing something about our scandalous budget deficit.

And every year members of the administration have had to admit that once again . . . guess what? There has been a slight miscalculation in the size of the deficit. In August of 1986, they had to concede that, no, it wasn't going to come out at $180 billion, as they'd said the year before, or even at $202 billion, as they'd recalculated it the previous February. Instead, the deficit would set another all-time record of $230 billion. Their original estimate was off by only $50 billion, that's all.

This is one game anyone can play. Start by underestimating the deficit from the outset. Pretend that some of those Medicare or defense costs won't show up in the next twelve months. Pretend tax receipts will come out bigger than they really will. Make the deficit look small enough to stave off a taxpayer revolt, and over the next nine or ten months, while hardly anybody is looking, quietly stick in the $10 billion here and the $15 billion there that you knew were coming due all along. It's easy, right?

The politicians 'fess up during the dog days of August, when most people are either at the beach or too busy worrying about next year's budget to notice that this year's budget somehow went wildly over everybody's expectations.

But these guys are wrong every year and the taxpayers are getting flimflammed. If you don't believe that, just look at the Reagan administration's track record at predicting budget deficits. Add up all its miscalculations between 1982 and 1986, and the overruns alone come to $278 billion. This administration added over $1 trillion to our national debt in those years, and more than one quarter of that amount came as a surprise.

The fact is, back in 1981, when the master plan was written, the administration people said they would balance the budget by 1985, and that in 1986 we'd actually have a budget surplus of $28 billion.

It's crazy. They've all been lining up for urine tests at the White House to prove they're not on drugs. They should have taken them back in the days when they were high on all this Reaganomics stuff. The tests would have definitely been positive.

When Congress passed the Gramm-Rudman law in 1985, Washington crowed that it had finally solved the deficit problem. That measure set a timetable ordering the budget balanced by 1991. Gramm-Rudman was going to be the panacea.

Well, no one can argue with the goal, but the budget-balancing law hasn't got a prayer. First of all, the methodology is unworkable. In a nutshell, the law says that we'll legislate good business practice by turning over the whole mess to some computers and a handful of clerks. But that's a meat-cleaver approach. We sent our representatives down there to choose among priorities. Instead, our peerless leaders decided to "automate" the process by putting some robots in charge.

Now, I know something about robots. In recent years I've bought hundreds of them. We use robots in our plants all the time for things like painting, welding, and heavy lifting. We let them do the dirty work. But we never use them for setting priorities or making decisions. We're old-fashioned. We still use people.

You see, Gramm-Rudman, being a robot, can be programmed to take $20,000 out of the budget, but it can't distinguish between $20,000 saved at the Pentagon on a spare part and $20,000 saved by laying off an FBI agent fighting crime. To a robot, all things are created equal. Even our smartest robots don't have a lick of common sense.

That's why we elect people, not robots. Whether they like it or not, our representatives have to be the ones to set priorities, make decisions, and then stand up and be counted when things get tough.

And the toughest calls should go right to the top.

In any event, to give you an idea of how much of a panacea Gramm-Rudman has been so far, the law specified that the deficit was to shrink to $144 billion in 1987. Well, the government brought it in at $158.4 billion.

Instead of getting mad at the administration, Congress simply revised the law and pushed the future limits skyward. Now the deficit doesn't have to be eliminated until 1993. I'll believe that when I see it.

In a democracy, of course, the final blame rests with "We the People." It's our fault that the deficit went to the moon. We're winking, too, because we keep listening to what the politicians are promising us for next year without holding them accountable now for what they promised us last year.

Politicians get elected by giving us goodies, not by taking them away. As Walter Mondale learned the hard way, asking for sacrifice is political suicide. Remember, he got creamed in forty-nine out of fifty states. So we can't get too mad at the politicians for letting the deficits run wild. Suicide is a lot to ask of anybody.

And that leaves nobody to fault but ourselves. The people in Washington live by polls. They don't lead public opinion; they follow it. The day the pollsters report that a majority of Americans are willing to sacrifice in order to turn our budget scandal around, you'll see it disappear.

In a democracy, that's the way it finally has to happen. The people have to lead their leaders. Main Street has to tell Pennsylvania Avenue: "That's it, boys. Your plan didn't work. Get a new one." We need a shareholders' revolt to send a message so clear that the politicians can't talk their way around it.

I'm convinced that if Americans really understood how deep in the hole they are, and just what they're doing to their kids' future, they would not only accept sacrifice, they'd demand it.

In business, the budget-cutting process is fairly simple. You call all your key people into a room and say, "Okay, fellas, sales are down and costs are up, so we have to take ten percent out of the budget starting right now. Come back tomorrow and tell me what you're going to give up."

No appeals. No debates. Everybody just goes back to his office and starts making a list.

That technique works fine in a typical corporation. The guy at the top can dictate, because he's not busy running for sheriff. That's why many of us who head big companies get the reputation of being tough guys. When it comes to budget-cutting, we've got to be tough. Consensus, or volunteering, is fine in some areas, but if you wait for a manager to volunteer a cut in his budget, you're going to be an old man.

But America isn't the typical company; it's a democracy. It's accountable to the people—so the process has to start with us.

Maybe we should all make a list of the most important things we get from the government. Protection from the Russians. A Social Security check. Medicare. A college loan. Cancer research. A mortgage deduction for the cottage at the beach.

Then we should ask: "Am I willing to cut something out to balance the budget?" I'm sure every congressman and senator gets thousands of letters with advice from their constituents (that's you!) telling them how to cut the budget. But I doubt that many of them say, "Start with me."

By the way, there are many people in our society who can't make up such a list. I'm talking about the disadvantaged—the handicapped, those living in poverty, abused children, the people who really need our support. It's not fair to pick on programs designed to help them. But that still leaves the rest of us.

Right now, entitlements represent 46 percent of federal spending, so there's got to be a lot of fat. Let's look at just one example. In 1986, more than 2,500 Americans with taxable income of at least $1 million received Social Security checks. Do those people really need the extra bucks? Hardly. So why shouldn't there be some sort of means test? If we stopped mailing Social Security payments to everyone making $50,000 or more, we'd instantly save $6 billion.

Social Security was invented as a safety net for the needy, not as mad money for the well-to-do. Incidentally, my mother is fairly well off and still gets a Social Security check of $480 a month. In fact, she just got a cost-of-living adjustment to $503 a month. That makes no more sense to her than it does to me. The basic problem, of course, is no one wants to shoot Santa Claus.

But the budget won't get cut until we tell the politicians to do it. They'll follow—they always have, once they hear our voices loud and clear.

In any budget process, when you've wrung out as much as you can on the cost side, then comes the heavy part. You've got to work the revenue side of the street, folks. Every business has to. After you cut and cut until there's hardly anything left, you have to go out and ring the cash register. You have to sell something.

So it's time we get serious and face facts; cuts alone won't do it. We need some revenue too. And that is precisely the issue no politician wants to face. It takes no courage whatsoever to stand up and say "I'll cut your taxes." Even robots can do that. But it takes a lot of guts to suggest new taxes.

At least in public. Privately, however, the worst-kept secret in Washington is that taxes are going up. It's only a question of when. Politicians are whispering about new taxes like little boys telling dirty jokes.

I'd like to see more tax on consumption and less on what provides all the capital and the jobs. Why not establish a national sales tax? It raises money like nothing else. Make exceptions for people who really can't swing it. If you make less than $15,000 a year, you're excused. But everyone else pays. Because we've got to stop dunning people for saving and start dunning them for spending.

The consumption tax should be graduated, as well, just like an income tax. Maybe 1 percent for a purchase up to $10,000 and 2 percent up to $100,000. However, if you want to buy a yacht, go ahead, but you're going to pay through the nose. As far as I can tell, nobody has to have a million-dollar yacht. It's not like a pair of pants or a dress. So why not slap a sales tax of 3 percent on that yacht? It may not stop you from buying it, but the government will rake in $30,000 in sales tax right off the top.

If we're really going to whittle down the deficit, we've also got to change the income tax code. But wait a minute—we just did that in 1986! (The first big change in seventy-four years.) True, but we screwed up.

A tax code should do three things: It should be fair to everybody; it should raise enough money to pay the bills; and it should help the country compete. Well, we spent a whole year on tax reform and got only one out of the three right.

I'll grant you, the new law is fairer. But since it had to be "revenue neutral," it didn't raise a dime against the deficit. And it shifted about $120 billion directly onto the backs of American business. So we got less competitive. Hell, Toyota made out better than Chrysler!

Every other country writes tax laws that encourage exports. Ours do nothing to encourage exports. I wanted to start exporting a few cars to Europe last fall. But guess what? I can make $1,100 to $1,200 more per car if I ship them from our plants in Canada and Mexico instead of from Michigan, Illinois, or Missouri—purely from tax savings! And that's enough to make or break the deal.

I've never favored all the personal income tax reductions we've seen over the years. Take me as an example. Just twenty years ago, my marginal tax rate was 90 percent. It went from 90 to 70 to 50 to 38.5 and now to 28. Believe me, I don't need all this money. Why doesn't the government freeze me at 35 or 38?

I do think 90 percent was a little steep; it may have destroyed my desire to work harder and make more money. There's not much of an incentive when beyond a certain level you get 10 cents on the dollar while the government pockets 90. Somewhere beyond a fifty-

fifty split you get a kick out of it. If I take home more money a day than my government, I feel I'm ahead of the game. Nothing wrong with that. After all, I'm the guy working for the money.

When we were going broke at Chrysler we didn't walk around handing out pay raises. We handed out pink slips. And we gave everybody who stayed with us the honor of taking a pay cut. That's what you do when you're going broke. You ask for sacrifice; you don't pass out goodies like tax reductions.

And so I'm willing to pay more taxes. I can afford it. But what I cannot afford is to bury my kids under a dungheap of public debt that will smother them and their kids, and their grandkids—forever.

Ultimately, the people do have the power to throw the legislators out. We can tell them: Either do the tough things that need to be done to get that deficit down, or all the angels singing in chorus won't save you.

And so, if nothing else, every one of us ought to go out and vote. With the exception of Switzerland, we have the lowest turnout of any civilized nation in our national elections. Good old free America. We believe in the Constitution and representation. And yet in recent years, only about half of us bother to vote in presidential elections—and even fewer in nonpresidential ones. Every time I see those figures, they jolt me. So maybe you'd better figure out what you're here for. You'd better exercise the responsibility for your freedom or you're going to lose it. In fact, you're losing it now.

I think we're going to see some action once Americans realize whom we owe this mountain of money to. A lot of people still say: "Don't worry about the national debt—after all, we owe the money to ourselves, so who's going to foreclose on us anyway?"

Well, guess what? The annual interest on that debt is the one thing we can't renege on. And unfortunately, we don't owe that to ourselves.

Here's the clincher. The people who are making those funds available are foreigners, notably the Japanese. We got hooked on Japan's cars and Sony VCRs. Now we're hooked on their money.

Japan already holds about $60 billion in U.S. notes and bonds, and these loans to us are growing so swiftly that it's estimated the total will reach $500 billion in the next five or six years. That's bondage, Japanese-style.

In 1985, in order to feed our debt monster, America became a debtor nation for the first time since World War I. And we didn't just slide into debtor-nation status; we dived in head-first. When we do something, we do it big. By rocketing past Brazil and Mexico, we now wear the crown of the world's largest debtor nation.

In effect, we're paying Japan reparations in reverse. When you win a real war, you get reparations from the countries you clobbered. When you win an economic war, you get some reparations too. This time the winner of the economic war is Japan. And they want to exact tribute from us. So a part of our wealth every morning goes to Japan in an envelope.

If we're such a rich nation, why have we had to go overseas to borrow money? Well, we don't save much in this country. We don't save enough to give the banks what they need to lend. Americans are great consumers, but they're lousy savers. I guess nobody reads *Poor Richard's Almanac* anymore—the part about "a penny saved is a penny earned."

Instead, we've become credit-card junkies. We're charging everything. It's hard to believe, but we hold roughly 70 percent of the world's 1 billion credit cards. There are now 105 million Americans with a total of 731 million credit cards stashed in their wallets. That's 7 cards per person! And nobody, I've noticed, leaves home without them.

Not only that, we couldn't care less about a little conservation. I think the best example of this consumption binge we're on is the way we've gone back to trying to consume the world's oil supply as fast as we can. Big V-8's are in again. You'd think, after standing in line like idiots twice in a decade to get our cars filled at the pumps, that a conservation ethic would have built up. Forget it.

We're right back in the 1960s—joyriding on cheap gas again. We just won't learn from history.

The import share of our oil is higher than it was just before the first oil embargo. And the Reagan administration has tossed federal fuel-economy standards for cars out the window.

I don't know how we got so blind so fast after all we went through. Do you remember those lines around the block, that feeling that someone out there had us by the short hairs? We cannot compete without a secure and independent energy source, but we've been capping our wells—and the Oil Patch has been bleeding the way the Rust Belt was a few years ago.

Because of our addiction to cheap gas, we've let ourselves get kicked in the head not once but twice. Now we're going to give the people who did it a chance for a hat trick.

Nobody wants to face the facts. We are spending ourselves silly. We're not paying our way. We're all looking for the free lunch. We want it all, and we want it now. And I'm not talking only about the so-called yuppie generation. I'm talking about all of us—many of whom are old enough to know better.

Isn't it common sense, after all, to save a little for a rainy day? At Chrysler we have a rainy-day fund. You know what it is? It's $2 billion under the mattress. All most Americans have under their mattresses is dust.

If we refuse to save but keep borrowing billions upon billions in foreign capital, we're going to find ourselves somewhere between a crisis and a catastrophe. And I'll tell you why.

Some people have the idea that all this foreign money is coming here because we're a so-called "safe haven." Well, that's becoming a myth. That money is coming to the good old U.S. of A. because we're bidding for it. Our interest rates are high. We've even changed some of our tax laws to make it easier for foreigners to invest here.

But now we're becoming too dependent on foreign money. Now we have to keep interest rates high in order to keep that foreign capital coming in, and to be sure that the money already here doesn't go home. Because if that money suddenly goes back where it came from, then we're up the creek. We won't be able to finance our debt. We could, of course, turn on the presses and print up some more money and put it out there. The only trouble is, that makes your currency worthless, and then inflation has a nasty habit of rearing its head.

This dependency is like a drug habit. You have to keep taking more and more. You can't cut back or you get sick. Stop altogether and you'll see the worst case of cold turkey ever. I mean, the whole world would go into convulsions.

Now we're even taking money out of countries that need it for their own development. Here's a shocker I didn't know: 22 percent of the foreign investment in this country comes from Latin America, mostly from private investors. That's right, from countries like Brazil and Argentina and Mexico that are strapped by their own foreign debts. While their governments have been busy borrowing money from us, we've returned the favor by borrowing from their wealthy individuals and corporations.

We're taking bread off somebody's table. Something's wrong here. It really is. Why is the richest country in the world siphoning off capital from underdeveloped countries to feed its own debt? And what will happen when those investors wise up and call in all their notes?

I started talking about the deficit in mid-1982, when it hit $100 billion. And I've kept talking about it and talking about it until I've grown hoarse.

I don't know who's on first anymore. I think I have a better than average I.Q. I've even got a master's degree. I've been in business forty years, and yet I'm getting a little frustrated trying to figure it out. I always thought that reasonable people finally came to their senses.

After all these years, it's unbelievable. The government keeps offering up the same old line: Don't worry, we'll grow our way out of these problems—as if they were acne or something. Somehow, some way, it'll turn out all right.

Tell me how.

Thomas Jefferson once said, "Don't appropriate money if you don't have a way to cover it." He said it was almost as bad as taxation without representation. That was his message 160 years ago. Now nobody wants to hear about it.

One thing is for sure: If we don't attack this deficit soon, we're headed for a real depression. Then we're going to see a lot of entitlements go out the window. We're going to see caps put on

Social Security. Roads and bridges aren't going to be kept up. Already, when you drive down some of our highways the potholes are so big you could lose your car. In New York, the bridges are actually falling into the rivers.

It's high time to batten down the hatches. And I can tell you very simply how to start. All you have to do is to turn off the TV for a minute and say: "Okay, kids, come in here. I just read this chapter, and I'm worried. Give me your credit cards. I'm cutting them in half. No credit cards for six months. We're going into hock too much. From now on, we'll pay as we go."

In sex advice columns, the adviser always says: "You must learn to say no to sex." Well, can't we learn to say no to spending money? Let's ask Congress: Can't you just plain say no to some of those spending bills?

I thought last year's stock market crash had at least some silver lining to it, because almost immediately the administration and Congress announced they were getting together to cut the federal deficit. But my optimism sank lower than the Dow Jones average when I saw the amount they had in mind: $23 billion.

That's $23 billion out of a federal budget of $1 trillion, or a whopping 2.3 percent cut. Talk about a finger in the dike!

I've got another number for them to consider: $100 billion. That should be the deficit reduction goal if they really want to convince the world they're serious about getting our house in order.

Six years ago, when the deficit was only $120 billion, it became obvious to me and a lot of others that our debt was getting out of control. When I was invited to the White House to talk about it, I proposed a plan to cut the deficit in half, with the pain spread equally between spending cuts and new revenues. I called it the "four 15's" because it involved taking $15 billion out of defense and $15 billion out of middle-class entitlements, for total spending cuts of $30 billion; then increasing revenues the same amount with a $15 billion surtax on imported oil and a $15 billion gas tax. (That was 15 cents at the pump, because every penny would have brought in a billion bucks.)

The administration didn't like the idea at all, because their pollsters told them Americans weren't in the mood for sacrifice. So the red ink kept flowing until October 19, 1987—Black Monday—when the bubble burst and the stock market dropped 508 points.

Now, at last, they understand that the budget deficit must be cut. But I'm afraid the patient is a lot more constipated after six years, so the dose of castor oil has to be increased. The $60 billion deficit cuts of six years ago need to be $100 billion today, and the "four 15's" of 1982 have to become the "four 25's" of 1988.

To make it politically salable, half should still come out of spending and half from revenues; that means $25 billion from defense and $25 billion from entitlements. On the revenue side, the penny on a gallon would still bring in a billion bucks, but now we'll need a quarter instead of 15 cents at the pump. And a national sales tax of about 1 percent on everything—again excepting food, medicine, and utilities—would contribute the final $25 billion. Six years ago I favored getting the last part of the revenue from an oil import fee, but today our revenue needs are too great to put all the burden on energy. A national sales tax would spread the pain around a little more, and it's still small enough that it wouldn't hurt state governments, who rely on sales taxes for a major part of their funding.

It's a nice formula, but there's one catch to it. The offer's only good for two years. Because without corrective action now, the budget deficit is going to be so high that my next plan will have to be called "four 50's"—that's if I'm not too depressed to keep trying.

The time has come for some fiscal responsibility. You see, it's all very simple. We have so much debt that there are only a few broad options available to us.

Option one is to default. Just not pay up. We'll simply say: "Oh, the hell with it. We're going to stiff everybody." Of course that's unthinkable. Maybe it's an option in South America, but not here.

Option two is that we could swap all the IOUs for our land and businesses. We'll offer Japan a few more skyscrapers and California, give them a state or two. But there's a limit to that. Geez—they already own most of Hawaii and a good chunk of California!

Option three is that we could start the printing presses and inflate the currency. Then we could pay back the debt at 50 cents on the dollar and miss the guillotine that way. But that cure is worse than the disease—remember Germany in the '20s?

Option four is that we cut our standard of living to undersell the competition. My economics department tells me that if we cut our standard of living by roughly 10 percent a year for 10 years, we could possibly succeed in getting rid of the deficit. I think the politicians in Washington would find that a really lousy plank to run on.

Those are the alternatives. Ah, there's one more. Start patiently to pay the price now. Regroup and get serious about competing through a combination of greater American productivity and smarter American economic policies—so that we can export goods and services to pay off the debt.

I don't know about you, but the last one is the only one of those options that makes any sense to me. But we'd better get cracking, because the sand in the hourglass is almost gone.

XII
THE FOOD CRISIS

There's a billboard in a little farm town in the Midwest that says: IF YOU COMPLAIN ABOUT FARMERS, DON'T TALK WITH YOUR MOUTH FULL. Well, I'm not chewing on a thing, so I'd like to say a few words about the plight of our farmers.

I got a letter recently from a farmer in Nebraska, one of dozens like it that I've received in the last couple of years. He said he was gazing out the window and wondering why the world was crashing in on him. "If something isn't done soon," he wrote, "this family farm of a hundred years is going down the tubes."

A farmer's wife wrote me from Iowa: "The decaying farm economy has taken its toll. My husband's health is declining because of a diabetic condition and the severe stress of the times. We have two sons, aged twenty-five and twenty-two, who have worked hard in the business since each of them was fourteen years old. They're willing to take on the business responsibilities but lack adequate finances. We just don't know what to do."

These are only two of the many people who are crying out for help from America's heartland. When you hear from people who are suffering, it gives you pause. They're part of families who have worked the land for three or four generations. They've always awakened at four, milked the cows at five, worked through every season, rain or shine, year after year. Some years were worse than others, but somehow they managed to survive. Now they're slaving just as hard as ever and yet they're dying. Many farmers are hurting so bad that they're feeding their cows apples and stale bread. Something's wrong. Really wrong.

During the recent prosperity in this country, the auto business made a comeback and a number of other industries revived—or at least did better than they had in 1980 and 1981—and yet farmers continued to drop by the wayside. And we haven't even hit a recession yet. It reminds me of comedian Joe E. Lewis's crack: "Hell, it's easy to go broke during a depression. *I* went broke during a boom."

Well, it's no joke to farmers. They've been going broke during one of our most euphoric booms.

Something's out of kilter here. The government's paying farmers huge sums of money not to plant their crops, while people are starving on the streets of our big cities. The legislators in Washington are doing so much talking about the problem, they're probably working up hunger pangs.

If we can't construct a sensible farm policy instead of telling farmers to plant one year and not the next, then what can we plan? I've been watching the decline of the farms, and frankly, I'm getting a little scared. We take farmers for granted, but they're so important we can't afford to mess around with them. As William Jennings Bryan once said: "Burn down your cities and leave our farms, and your cities will spring up again as if by magic; but destroy our farms and the grass will grow in the streets of every city in the country."

I gave a speech in late 1987 to the Future Farmers of America, those bright-eyed, bushy-tailed young kids who hope to be tilling the land in the coming years. I tried my best to be upbeat and to hold out a ray of hope for them. But all the while, in the back of my mind I kept wondering whether some day their logo would have to change to the Former Farmers of America.

Young farmers are perplexed, and I don't blame them. The America they see today is a lot different from the America their parents grew up in. Some of their parents can probably remember when 20 percent of all Americans were farmers. Today it's under 3 percent, and there's some question about whether we even need that many. Farmland is valued at its lowest level in almost twenty years, at about $600 per acre, a decline of 33 percent over the last four years. According to Department of Agriculture figures, a farm went out of business every ten minutes in 1985. In 1986, the rate rose to one every seven minutes. By the end of the century, some estimates are that at least 200,000 more farmers will have thrown in the towel; other estimates run as high as a million.

I'll bet those young farmers' parents can remember when the word that was used to describe their profession was *agriculture*. Now it's *agribusiness*. They've seen the computer replace the old *Farmer's Almanac*. Today they don't turn on the radio in the morning just to find out what the weather will be in Iowa; they also have to know what the value of the dollar is in Tokyo.

Money markets are now affecting currency rates, and currency rates are affecting the ability to ship a bushel of wheat out of Kansas. The farmer doesn't know why there's suddenly no market for his wheat in Europe. There was a good market just a year ago. Sure, but the currency changed.

Even a big-city guy like me has seen the difference. I've been in marketing all my life, but I always thought it was something you did for cars or toothpaste. I nearly fell out of my chair the first time I saw a television commercial for eggs.

I can also remember the newspaper stories just fifteen or twenty years ago about the starving masses overseas whose only hope was the American farmer. Well, the Green Revolution of the '60s and '70s has worked some remarkable magic. Today, so many more countries are self-sufficient. In fact, some of those that were starving are now exporting food.

When I was growing up, I was always being told about the hungry millions in China. If my father spied a few scraps on my dinner plate, he'd scold me that those morsels could feed ten Chinese. Now China produces all the grain it needs. In fact, it's selling its surplus corn to Japan more cheaply than our farmers can. Thailand is shipping more than twice as much rice as the United States. Indonesia, the largest importer of rice just five years ago, now has plenty from its own crops. In 1986, India grew so much grain that it didn't have enough room in its storage facilities to hold it all. And seventeen years ago a famous rock concert was staged to raise money for the malnourished of Bangladesh; these days, Bangladesh is self-sufficient.

Believe it or not, for a short time in 1986 the United States actually ran a slight trade deficit in agriculture for the first time ever. Alarm bells should have gone off all over the place when that happened, because if there's one American who is better equipped than any of the rest of us to compete in the world market it's the American farmer.

But you can't compete with the guy who won't let you in the door. Nothing is more protected around the world than agriculture. Every government I know of—including our own—protects its farmers from too much foreign competition.

That's why the Japanese pay six times the world price for their own rice: to make sure Japanese farmers keep growing it. I guess it's a result of their island mentality. They figure that if some foreign power ever tried blockading them, at least they would always have rice to eat. The Japanese can get all the beautiful, inexpensive American rice that they want—but they're not interested.

Obviously, competitiveness involves a lot of things: hard work, brains, ingenuity, technology, capital investment, education—the list goes on and on. And yet you can do everything right, but if you don't have the right government policies to back you, all your brilliance will still leave you at the starting gate.

I don't know how many times I've listened to that old broken record that says that all the American automobile industry has to do to beat the imports is to get more productive. Just build cars more efficiently and you'll drive the other guys back into the sea—that's the lecture I hear all the time.

And when I do, I always point to the American farmer and say: "Wait a minute. He's the most efficient and productive farmer in the whole world. If productivity is the simple answer, how come so many farmers are losing their land, even while many draw billions and billions in government subsidies every year?"

Whatever the farmers' problem, it's not that they aren't productive enough. (They're too productive.) It's not because they don't have the technology. (They are the most mechanized in the world.) And it's not because they haven't made the investment. (These guys literally bet the farm every single year.)

We're not competitive in a lot of things in this country, but farming ain't one of them. Even though we've lost certain advantages to some of the European farms, we still swamp them by any number you want to use. So the guys in Washington can't tell the farmers: "Go back and plow your fields and stop being crybabies."

Farmers are producing so much food that we're running out of ideas about where to put it. In Dubuque they've been using caves to stuff the corn in. Along the Missouri River they've been tying up barges and filling them with produce. Clearly we don't need more food. In fact, billions of dollars in this country are being spent on diets.

I read recently about a California dairy farmer who was so productive that the government paid him $8 million to slaughter his herd. At those rates, I don't know how productive we can afford to get! Unless, of course, we can start opening up some foreign markets.

I don't pretend to be an expert on farm policy, but speaking as an outsider looking in, there are some things I just don't understand.

I don't understand, for example, why foreign owners of American farmland are collecting federal subsidies for crops we can't export because of all the foreign trade barriers. In 1986, $2.2 million of those American subsidies went to a group that included the Crown Prince of Liechtenstein. Now, I don't know the Crown Prince of Liechtenstein. He's probably a nice guy. But I don't think that he's exactly down and out. And I have to wonder why he and his friends collected $2.2 million from the U.S. Treasury while I was watching American farms being auctioned off on the six o'clock news a couple of nights each week.

I've heard about this farm support system all my life, and I've always been befuddled by it. Even my father collected a bit of subsidy money once. He owned a plot of four hundred acres in Pennsylvania, on which he eventually developed houses, and for one short period he got paid for not planting corn there. When the check arrived, he dashed off a letter to Washington asking why he was being paid to keep the land fallow. He didn't understand it. (Times were pretty tough, though, so he kept the money. He wasn't *that* indignant.)

Like my father, the guy on the street finds it difficult to comprehend why we pay somebody for not doing something. It's hardly the American way.

The irony really hit me a couple of years ago when I saw the amount that the government was shelling out. "That number's got to be wrong," I said to Don Hilty, my economist at Chrysler. I couldn't believe it was so big. Here I had tried to get $1.2 billion to keep our factories going, with people yelling at me from all sides that the government shouldn't bail out people who can't make a go of it. And yet in one twelve-month period, ending in September 1986, we spent $26 billion paying farmers to sit tight. And it still wasn't enough to keep a lot of farmers from going belly-up.

The bill for the federal farm programs over the three years ending September 1988 will total at least $72 billion. That's almost 40 percent more than the cost projected in late 1985, when the current law was passed. We're pumping money into farms at a record clip, with no end in sight. I have to wonder: Once the guy gets his bundle of money, is he a farmer anymore? Sounds more like a lottery winner.

Some of the federal subsidies directly contradict one another. For instance, dairy farmers received $2.4 billion last year to encourage production and $1.1 billion to discourage production and another $400 million to compensate those who sold their cattle to slaughterhouses. That's nearly $4 billion to both discourage and encourage production. I suppose this makes sense to somebody, but it doesn't to me.

I'm not sure what the farm plan should be. I don't even sell tractors. It's not a simple problem, and so solutions to it can't be simple.

After all, farmers have always ridden an economic roller coaster.

Actually, America had its first farm crisis, in the form of Shays' Rebellion, in 1786. We've been fighting a food surplus now for more than a hundred years. A couple of world wars, the Korean War, and the Vietnam War helped to reduce the surplus, and a welter of farm programs have tried to correct what the wars didn't. Farm programs keep coming and going, every one of them complete with loopholes big enough to drive a combine through. Clearly, the current plan isn't getting the job done. It needs an overhauling from top to bottom—that much I do know.

I also know that there seem to be some striking similarities between farming and the auto business. Both of them face worldwide overcapacity. Both of them face unprecedented global competition. Both of them are heavily affected by acts in Washington. Interest rates, trade policies, energy pronouncements—each translates into costs for both industries.

Like our business, farmers bear some responsibility for their predicament. Chrysler and the farm industry took a similar path during periods of expansion—they overdosed on debt. In the late '60s and early '70s, when the car companies were riding high, Chrysler expanded to become a worldwide company; it overextended itself and made some management mistakes. The farm industry did the same thing in the late '70s. When exports went through the roof, farmers expanded and took on more debt than they could handle. Then, when things collapsed during the recession in the early '80s, they were stuck holding the tab.

There are a lot of farmers who speculated in land and interest rates, paying too much. Now, with the prices they're getting, they can't make the payments and they're losing their farms. They made a bad deal, and as a result there are far too many farmers in hock up to their ears with the banks. Farmers have probably figured out by now that when you lose control of your debt, you lose control, period. Today, the farmer has to go see his banker before he even decides what to plant.

So I sympathize with farmers. I've been there. In hard times you wonder if you'll ever see daybreak again. The answer, though, isn't to keep pouring billions of dollars into farm subsidies year after year.

Sooner or later the taxpayers are going to demand that we stop picking their pockets. Especially since the more we spend, the worse the problem seems to get.

Farmers, however, wield so much political clout that every senator and every representative considers himself a friend of the farmer, especially the small family farmer. And that's the main reason subsidies continue to grow, even when the plants don't. That's also the main reason nobody in Washington wants to face a harsh reality. But common sense dictates that at least some of the less efficient farmers have to close down and find another line of work.

I know that sounds radical. I know no farmer wants to give up land that's been in his family for generations and a life that's more rich and meaningful than any nine-to-five job. But there's no reason to worry that once farmers leave their land they'll simply fade into oblivion. No one I know works as hard as a farmer, and that's an ethic that any employer is bound to find pretty attractive. Quite a few states already boast retraining programs to enable those who give up farming to pick up a new trade.

So maybe 20 percent of our farms have to close down, just as 20 percent of our auto factories had to. Of course, that's a tough thing to tell the 20 percent that have to go, but something's got to give or one day all of our farmers will be gone.

Perhaps some farmers could survive if they learned to diversify their fields. Governor Michael Dukakis of Massachusetts has pushed that approach. Farmers growing soybeans, he's suggested, might take a shot at flowers, blueberries, and Belgian endive. The idea has been labeled "yuppie agriculture," and the name alone probably dooms it, but I don't know that it's such a crazy thought. Of course, a guy in Kansas could probably say it's easy for a guy in Massachusetts to make it sound so simple.

One obvious way to clear up the farm crisis quickly is to get rid of subsidies altogether. Just turn off the money spigot and leave farmers no choice but to grow only the amount of food that people want and have the money to pay for. But that's a solution that would put so many farmers out of work so fast that it would make today's bankruptcies look like child's play. It's too cruel an answer. Some form of monetary assistance has to continue or else we'll be in even deeper trouble. If the government leaves farmers completely to their own resources—makes it a game of survival of the fittest—we might end up with a couple of superfarms. We might have Cargill producing all of our wheat and Seabrook Farms growing all of the vegetables. There's a word for that: monopoly. Then the government would probably have to step in and break them up anyway, just to inject a little more competition.

I do think we have to weed out some farms—but let's not get down to two. And let's do some planning, for once, so that the weeding is done in a gradual way that's as painless as possible.

At the same time that we try to scale back our food output, we ought to figure out a way to use some of the excess. Even though the world seems to be growing enough food for everyone, that food's not finding its way to everyone's table. Sure, we've come a long way from eighteenth-century France, when kids were left along the roads because their parents were unable to feed them. Yet, although Bangladesh and China may be able to take care of themselves, we've still got plenty of empty stomachs in countries like Angola, Mozambique, and Chad.

According to the World Bank, between 350 million and 700 million people worldwide don't get enough nourishment each day to be able to work or avoid stunted growth. And lest you think that the bellies are rumbling only overseas, take a look at what's happening to us. Not long ago I did a commercial for the street people of Detroit— the poor who are curled up in the gutters—to try to solicit money for rescue missions for them. I was horrified at the dimensions of the problem in my own city.

I know that our government has handed out groceries from time to time—it gave away some butter that was getting rancid and it distributed a bunch of cheese—but not often. I have a lot of trouble understanding why the silos are full, we're paying guys not to plant any more, and yet we can't seem to figure out a way to feed the poor.

The political answer always seems to be that if you're not careful you'll end up giving away food to people who would otherwise buy it, and thus you'd drive still more farmers out of business. George

McGovern, nevertheless, was convinced that we could set up an International Food for Peace program that would allow surplus food to be funneled to needy people without undermining the commercial markets. Something has to be done to end the outrage of too much food coupled with too much starvation.

We're lucky that we still have those young, bright-eyed men and women who want nothing more out of life than to farm the land. The business is so chancy now that I know I wouldn't ever want to be a full-time farmer. I have a hell of a time with my little Italian grape and olive operation. Last year I had to cope with a week of terrible rain and then I used the wrong kind of fertilizer, and the crop was lousy. I worried myself sick, and it was only a hobby.

Every year in this country, there seems to be some sort of agricultural calamity. Either the boll weevils eat the cotton crop or there's a drought in Georgia or a freeze kills all the tomatoes in California. It's hard enough running a business where I've got to gaze three years into the future and figure out people's taste in cars. At least I don't need to dread the weather report.

We assume there'll always be somebody out there willing to go into that risky environment and grow the food. In a real pinch, the average person thinks: *Well, I could start a victory garden in the backyard. I've got a little plot there. I'll just pick up some seed and do my own farming until things straighten out.* I don't think we realize what goes into putting food in our mouths. People know they can't grow everything, but they figure there's a canned-goods department in the supermarket and that whatever is packaged in a can somehow appears there magically. No matter what happens out in the fields of Iowa, they expect that they can always scrape by with canned food.

During the gas crisis, people lugged around ten-gallon cans in their trunk and stashed away hundred-gallon drums in their basements. I even put aside a couple of five-gallon cans in my house just in case. Everyone had taken gas for granted. We found out the hard way that there wasn't a limitless supply.

One morning we could wake up and find there was no longer a steel industry in this country, or we could wake up and find our car business gone. I've often heard the line "But we must have a machine tool industry and a steel industry for defense in case there's a war."

Well, if we wake up and there's no food left in the United States, we've lost our most important defense. We can't even wage peace without food. We can get by without new clothes for quite a while by stretching the old ones. But we can't stretch not eating for very long. If we can't lick our agriculture problem, we don't have to worry about any of our other worries. Once we stop eating, both our problems and we disappear. Think about that when you dig into your pot roast tonight. And please, don't talk with your mouth full.

XIII
THE SCHOOL CRISIS

When I was a kid, being a good student was very important to me. And in case I ever forgot just how important it was, I always had my father around to remind me.

But something's happened since the days when I was clapping erasers and washing blackboards. I'm the co-chairman of the Governor's Council for Jobs and Economic Development in Michigan. At one of our recent meetings we began talking about the state of education today. The discussion quickly got down to basics, with the people in the room tossing around some pretty frightening facts and figures.

Doug Fraser, one of the council members, said, "Do you know that the armed forces just did a study on the I.Q. levels of their people and found they were lower than they were ten years ago? After the great high-tech explosion in information and training, their I.Q.'s actually went down."

This comment immediately set off a game of Can You Top This?

"That's nothing," someone else chimed in. "Did you know that sixty percent of the people at the ninth-grade level can't even read properly?"

Then Martha Griffiths, our outspoken lieutenant governor, offered her perspective: "If you ask me, the monster is the tube. All the sitcoms and the other garbage on TV should have subtitles running across the bottom of the screen, as if the shows were in a foreign language. Because the kids never see the words. They get impressions of voices and situations, and they end up not being able to read or write."

Then she added, "At McDonald's, the reason they show pictures on the menu is because kids come in and have to point. Spelling *double cheeseburger with bacon* is not that easy."

The observation was all too true. International sign language came into being because languages were different throughout the world; NO SMOKING became a red sign everywhere, and SCHOOL CROSSING a yellow sign. Now we're forced to use sign language for our own kids—and they're supposed to be speaking English!

A country's competitiveness starts not on the factory floor or in the engineering lab. It starts in the classroom. We've got to get cracking on education—at all levels—or we'll get run over by the Far East. The situation is getting pretty bad. We're not progressing; in fact, we're going backwards.

I have to wonder about a country in which two out of three high school juniors don't know in which century the Civil War was fought. I have to wonder about a country in which one out of five high school kids believes the telephone was invented after 1950. I really have to wonder about a country in which 700,000 kids who graduated from high school in 1986 couldn't read their diplomas!

To make matters worse, an appalling number of kids aren't even bothering to complete their educations. Nearly a quarter of America's teenagers drop out of high school every year, and that means a million untrained workers are hurled into the job market. A recent study of young working people found that a fifth of them read below the eighth-grade level. That's almost as bad as not reading at all, because most of the material in workplaces is written for a ninth- to twelfth-grade level.

All too often, these workers wind up committing expensive errors. And so you hear about the steel-mill worker who once ordered $1 million worth of the wrong parts. Why? He couldn't read the instructions well enough. Or the insurance clerk who paid out $2,200 on a $22 settlement. Why? She didn't know how decimals worked.

I guess I shouldn't be surprised. The latest estimate is that about 27 million Americans are functionally illiterate. That's a shocking number in the late twentieth century. If this keeps up, I won't bother to write any more books—nobody will be able to read them.

Things are so bad that three out of four U.S. corporations—Chrysler is one of them—are forced to train new workers in basic reading, writing, and arithmetic. It's reckoned that the lost productivity is tacking on a whopping $25 billion a year in costs to American industry.

Obviously, everyone needs to wake up and get a lot more serious about education or we're going to be in a hopeless mess. The government sponsors a lot of programs—a farm program that goes up and down year to year, a defense program that is also as adjustable as a yo-yo. But there are some vital programs that no government should tamper with—except to improve them—and the educational system is one of them.

In 1986 the administration proposed cutting $5 billion out of education at the same time that the President told all of us that America had to get more competitive. Who's he kidding? As Derek Bok, the head of Harvard, put it: "If you think education is expensive, try ignorance."

To me, education is the price of admission into our democracy. Having a strong long-term educational program is the core of a good strong nation. It will take time to be developed, but it certainly shouldn't waver with the budget. I'm all for spending cuts, but that's a gamble we can't afford.

We also waste entirely too much time fussing about the peripheral things, such as whether we should start the school day with a prayer. Nobody seems able to decide what school prayer is or whether it's even okay to meditate in silence. But come on, will school prayer really make or break our future? Think about it. It's the kid pregnant in the ninth grade who's mortgaging her life. It's the crack being sold in the halls to kids who don't believe in this life, let alone an afterlife.

Don't get me wrong. It's not that I think our educational system is in a shambles. But I do feel it's crying out for a lot more attention. The standards of what constitutes good education have changed, because there's a new guy in town called Japan, who's really concentrating on his homework.

If you look at some of the most recent statistics on American education, they're mind-blowers.

Internationally, American students ranked no higher than tenth out of twenty industrialized and Third World nations in a 1985 assessment of math achievement. Guess who was number one? Japan.

American students spend 180 days a year in school. Japanese kids, including the half days they put in on Saturday, spend 240. Overall, by the time a Japanese student graduates from high school, he or she has been to school one to three years longer than an American student. As a result, the average Japanese high school graduate has the basic knowledge of the average American college graduate.

The Japanese are so fanatical about education that when a kid is sick his mother will often show up at school in his place and take notes for him. Can you imagine if the kid got the mumps just when they were starting differential calculus?

Some experts are now arguing that it's impossible for us ever to keep pace with the Japanese, because we spend sixty fewer days a year in school than they do, and so our performance will never be more than 70 percent of theirs.

I don't quite agree. My feeling is that the quality of an education depends more on how you introduce kids to the basics than on how many days you keep them at it. To me, the ten-year span from kindergarten through the ninth grade is the crucial period in the educational cycle. The Japanese think education should start practically the minute the umbilical cord is cut. Their kids begin school at three or four—and they even have to compete to get into kindergarten. I don't know how they evaluate them at that age; I guess with finger painting and Slinky contests.

Because I believe it's the ages from five to fifteen that really shape a person, I think this country puts too much emphasis on secondary schools. Japan does just the opposite. There, the tough years are grammar school and junior high. In fact, only 29 percent of Japanese students even go on to college, compared to 58 percent of Americans. But 90 percent of Japanese kids graduate from high school, while only 75 percent of our kids do. By the time a kid there has finished high school, according to the Japanese, his values are already formed.

And so the students who go to the University of Tokyo are an elite group. Anyone accepted there is really in—you almost never hear about any other colleges. But when Japanese companies are hiring, they don't seem to care much about how any of their prospective employees did in college. Or if they even went. The boss never bothers to look at the grade-point average—only the name of the school.

I don't know if we could, or should, de-emphasize college that much here. In Japan, after all, everyone in a corporation bands together for life and no one ever gets fired. As a result, the company is like a continuing education. There's no floating around. Nobody over there joins an auto company and then after three months decides: *I think I'll go into the bowling alley business.* There's no midlife crisis with the Japanese.

Although I don't think we should play down colleges as much as the Japanese have, we've gone a little overboard in the opposite direction. The attitude at a lot of our major corporations has been that if you didn't go to Harvard Business School and get an M.B.A., you're a misfit. When I was at Ford, in fact, the controller always carried around a tally of how many Harvard graduates we had. It was tucked safely into his front shirt pocket for easy access—not the sales figures, not the profits, but the M.B.A.'s from Harvard. The entire finance group seemed to believe that as long as you had a doctorate or a master's degree, you could make cars in your basement at home with pipe cleaners.

Fortunately, this attitude is abating somewhat. People are beginning to understand that the ultimate case study is not at Harvard; it's behind desks and under engines at Chrysler or Toyota. The message is slowly sinking in that the real way to learn is to go out and get your hands dirty.

After all, you can learn a lot more once you've settled into a career. The South Koreans are big promoters of this philosophy. Samsung, for instance, does a fantastic job with all its people. When anybody changes a job or is hired, he or she gets intensive in-house training. Even if you're a vice-president and get transferred to another job, you must go to the vice-presidential school. You're given a week to study why you got promoted and what you're going to do now that you've got a new job.

The Korean approach brings to mind a quip I heard recently: "A college education seldom hurts a man if he's willing to learn a little something after he graduates."

Unfortunately, the crisis in our schools goes beyond the amount of education that's offered. The question you've got to ask is: What kind of education is being served up? You can go to school every day of the year, but if all you're studying is beanbags, what good is it?

Japan, on the other hand, really keeps students focused on their studies, pumping fact after fact into them. And their society doesn't tolerate any distractions or outside indulgences. Most high school students in Japan don't drive, don't date, and don't help out with the dishes at home. They go straight to their room and hit the books. Just in case they're tempted to have a little fun, the teachers even give them homework to tackle over their forty-day summer break. In fact, I understand the instructors are so strict there that if a kid dyes his hair, he gets punished as severely as if he'd given the principal a hotfoot.

Meanwhile, American kids back home are watching an average of fifteen thousand hours of television between the ages of six and eighteen. That's two thousand hours more time than they spend in school. So, while the Japanese kids are boning up on economics and math, our kids are memorizing the names of the kids in *The Cosby Show* and the intricate plots of *Cheers* and *My Sister Sam*.

Every kid reaches a juncture when an inner voice says, "I have to decide whether I'm going to be a lawyer, an engineer, or a butterfly collector." When I went to school—I think it was in the ninth grade—we used to have a formal class called "guidance," which was intended to do exactly that: guide you toward a career path. The teacher would tell you about colleges and curricula, and then he'd lay out the facts: "Remember that to become a doctor, you don't just go to college for four years. You've got to go four more, and then eight more, and then you're on call day and night. You're going to see an awful lot of blood and pain and suffering. (And you can only play golf on Wednesday afternoons.) It's one tough job." Or he'd say, "Some of you probably like the outdoors. Well, let me tell you about the forest rangers."

After I finished the course, my guidance teacher called me in and said, "Okay, first you've got to think about what you're going to do in high school."

Like most kids, I replied, "I haven't a clue."

The teacher studied my record and told me: "You're good in math. You ought to follow it up." And so it was decided that I would take engineering and science in high school. I was also good in Latin, and I was right up there in English too. He just happened to focus on math. Sensing I wasn't sure what I wanted, he said, "If you're good at a lot of things, why not take something formal, such as engineering and mathematics, because if all else fails, those courses will at least discipline your mind."

"That makes sense," I figured. "I can always read history and do English on my own, or play the saxophone, but I guess I better have this background of scientific stuff."

So I went to high school and took engineering and math. When I got to my senior year, I said, "Well, nothing's changed. I still don't have a clue about what I want to do, so I guess I'll just go on to engineering school."

Off I went to Lehigh University to study mechanical engineering. At the end of my sophomore year, however, not only were the courses getting awfully tough, but I began to think about going into business. To do that, I knew I ought to have a broader mix of courses. I liked science, but I also wanted to know how it applied to making something work. To this day, I feel knowledge is useless if you can't apply it. That's what the Japanese say: "You guys do the inventing, but we take your inventions and we process them and apply them and make them into neat little things like VCRs and cars." My sentiments exactly—in 1943.

It turned out that I was pretty good in my business courses, and so when I left college, that's the career I pursued.

What I'm trying to say is, education shouldn't be highly specialized. In my day they tried to make you specialists. And, not knowing any better, I believed that was the smart way to go. I was one of those science snobs who looked down my nose at liberal arts majors. Why does anybody need to study liberal arts? What do those guys do anyway? Read history books and write poetry. It's ridiculous.

Well, I've finally realized what they were driving at. Now we have kids who are so specialized that they know all there is to know about one piece of the pie and nothing about the pie. They may have a lot of facts at their disposal, but it doesn't help them think. To figure out how to create a better future, you'd better know what went wrong in the past. Today I read more history books than I ever did.

One area of specialization that I'm particularly skeptical of is computers. In recent years, a lot of hoopla has surrounded the arrival of computers in the classroom. Frankly, I'm not so sure that the ability to work a computer is all that essential to the future of this world. After all, what is a computer? It's a sort of brain that you can nimbly call on with your fingers to obtain information. But what are you going to do with all that information once you get it?

Some of the little kids in my neighborhood are absolutely fantastic with computers. At twelve years of age, they're masters of the Macintosh. Day after day, they plug themselves into that machine as if it were a life-support system. But does all that time at the screen really teach them to think? I doubt it.

Now, I don't dispute the importance of computers. My secretary uses a word processor, so now she has time to do important things instead of just pecking away at a typewriter. Still, at Chrysler the cost of projects for computers has been soaring at an unbelievable rate. Naturally, the guys who insist on how much we need them are the very brains who like to use them—namely, the fellows in the controller's office.

They'll march in to see me and announce that if we spend $25 million to put in this new mainframe from IBM, all of our systems will be able to talk to one another.

"Great," I tell them, "but what does that do for the company? How do we get more productive and build better products?"

"Glad you asked," they reply. "This mainframe is going to save us twenty-two new employees, easy."

So we invest in the computers and in training personnel—millions of dollars, thousands of hours—but we never go back and check on whether we saved ourselves even one person, let alone the twenty-two. I've signed so many projects that by now I should have nobody left. Supposedly, the new system is feeding us more data and more scientific analysis of that data. Yet I don't know that our decisions—or our cars—get any better.

I'm already being swamped with far too much information. I am getting paralysis by analysis. Every time I mention something, I'm handed ten thousand printouts that spit out of a computer. My guys can give me a report on how many teenagers are buying cars costing over $8,000 on a Tuesday afternoon in Albuquerque. What I'm going to do with that information I haven't figured out yet.

Unfortunately, the computer can't make a decision for me. It can only serve up alternatives quickly. The experts call them scenarios, and sometimes I feel I'm getting scenarioed right off my feet.

Most Japanese kids are trained on a computer at an early age, but I don't think that's where their success comes from. They're smart and they're highly disciplined. They're more organized than we are. None of those traits has anything to do with computers.

There are some essentials that have to precede technical skills. One of the most important habits students ought to pick up from an education is the love of reading. You've got to read a diversity of things in order to have opinions. Then you've got to know how to communicate those opinions to your fellow man. And finally, you've got to learn to collect your thoughts and be able to write—at least modestly well. Some kids can't write a letter home to their parents, because they're unable to spell or put sentences together. The grammar and spelling of kids today just blows my mind. Their compositions read as if they were written by cavemen. You know, "Me Tarzan, you Jane."

I always tell my kids that they ought to read as much as they can. You don't have to be a voracious reader, but if a day passes and you've read nothing, then you've learned zero from that body of lore out there which has accumulated over the past two thousand years. You may have run into your secretary, who says, "It's a nice day, isn't it?" And that's the sum total of what you learned. When you think about it, you didn't even learn much about the weather.

Some people will read, but they won't really read. Because they won't remember a single fact from what they read. They skim, but nothing sticks.

I'll never forget something that happened to me when I was in third grade. We were studying the Greeks, and the teacher said, "Okay, who knows the name of Ulysses' dog?" I don't know why, but for some reason I alone remembered it. The teacher complimented me, and the other kids were in awe of me—for at least a day.

The name certainly didn't help me later on in life, because today I haven't the foggiest idea what the mutt's name is. But the fact that I knew something from my own reading, something that the other students missed, motivated me. From then on, I wanted to make sure that I read and retained even the trivial things. In business, they're often more important than the big ones.

A friend of mine named Bill Kanehann and I went through a time in high school when we studied vocabulary so much that we played a little game to keep ourselves sharp. The idea was that one of us would begin a conversation and try to use a particular word in a sentence. Then the other one had to change that word to a synonym. We'd keep this up until one of us ran out of synonyms.

A typical exchange would go:

"You're talkative."

"No, no, you're not talkative—you're garrulous."

"No, you're not garrulous—you're loquacious."

"No, you're not loquacious—you're verbose."

Bill was so terrific at it, it was infuriating. He'd never run out of words. Somehow, no matter how long we'd been at it, he'd always find one more word tucked in the back of his head. If nothing else, those early mental gymnastics have made me a whiz at crossword puzzles.

A few years ago, Bob McNamara's wife, Margie, got me involved with Reading Is Fundamental, an organization committed to teaching kids to read. There are few things I feel more strongly about. I did a commercial for the organization and a poster for all the libraries. Its message is: "If you want to get ahead, you better read, you better read, you better read." I believe it, too.

Easy enough to say, but to know what to read, you've got to have the right people directing you—and that means good teachers.

It's amazing how long this nation has debated something that shouldn't have to be debated at all: how much we pay our teachers. I happen to think teachers are so important that they should be paid a lot more than they are. It's not just a lot of malarkey that they're being shortchanged.

In fact, the average American teacher's salary of $26,700 a year is just below the $27,089 that the average postman gets. I like to get my mail on time as much as anyone, but I'm a little more concerned that my kids or the people I hire know where Nicaragua is on the map or how to read their income tax forms. (On second thought, that's asking too much of anyone!)

Even a chauffeur for the Board of Education in Detroit makes $1,600 more than a starting teacher. It's no wonder that a 1985 Harris poll found that one out of four U.S. teachers will flee the profession within five years because of the low pay and dismal working conditions.

Determining what's the right pay for a teacher isn't easy, but I guess you could do it the same way we set pay at Chrysler. Assign a degree of difficulty to a job and then decide the pay ranges. Within those ranges, try to stimulate employees to chin themselves up so that they perform to their fullest.

Sure, it raises a hard question. How important is the grade-school teacher compared to the college professor in dealing with our resources known as kids? I'm not sure who should decide that. But money motivates people. You must start there. Teachers have kids, too, and they like to take vacations and spend holidays with the family, just like the rest of us. Meanwhile, they teach long days and have to grade stacks of papers at night. It's no nine-to-five job.

In Japan, teachers are in the top 10 percent of wage earners. No wonder some of the best young people there become educators. In this country, the opposite is true. Only 12 percent of the country's graduating high school seniors now major in education, down from 20 percent in 1970. By 1990, there could be a shortage of 500,000 teachers. What will that mean for our kids—and for us?

Last year, the Kansas City school district set up six foreign-language schools to teach French, Spanish, and German to elementary-school kids. They scoured the entire United States for grade-school teachers qualified to teach French who were willing to come to Kansas City. After looking everywhere, they couldn't come up with a single one. Finally they had to import fourteen instructors from Belgium. And then, to top it off, the Immigration and Naturalization Service refused to let them in.

Things are so bad in our schools that studies show about half of the newly employed high school science and math teachers in the United States are unqualified to teach their subjects. In fact, the pool of science teachers in the United States is drying up so rapidly that the job is increasingly being handed over to the football coach. He may know something about immovable objects meeting irresistible forces, but I wouldn't trust him with mixing chemicals in the lab.

Teachers do spend an incredible amount of time with our kids. When you as a parent have a new child, you're going to spend a lot of hours with him—from zero to five years of age. But once he's hit five, the teacher he draws out of this school lottery will have more influence on him than you'll have. When he's at home, the kid is usually either sleeping, watching TV, or tormenting his sister; when the teacher reaches him in school, she's got him for six to eight uninterrupted hours.

Obviously, it's essential that you give your kids a good send-off at home and instill the right values in them. But those values can be distorted instantly by the kindergarten teacher or the high school teacher, not to mention some of the pointy-headed professors at our graduate schools.

As a teacher, you can have bad apples as students. But as a kid, you'd better not have too many bad apples as teachers, because their impact is so great. One bad apple leading one course with forty kids can produce twenty counterrevolutionaries faster than you can say "Down with capitalism."

I was lucky. I drew fantastic teachers one after the other. I seldom had a dud. Mine were always worrying about us, caring for us, being demanding of us—and disciplining the hell out of us. Classes were small, and there was a personal feeling about them. Our teachers were always trying to wring out the best from us. I'm convinced that everybody's got some genius at the core. The role of the teacher is to yank it out, and mine did their damnedest.

When I was in school, the teachers motivated us by turning drudgery into a competition. In my junior-high English class, for instance, we had to write a five-hundred-word essay that was due every Monday. We would get a gold star for an outstanding paper. During the course of that year, those students who received three gold stars hit the jackpot: They didn't have to write any more essays. You'd better believe everyone worked on those papers until their eyes were blurry and their fingers ached. I know I did more drafts then than I do today on my speeches and columns.

But those stars came mighty hard. Jimmy Leiby, one of my best friends, was the first to get three of them. It probably didn't hurt that his father was a newspaper reporter, so maybe it ran in the family. His success egged me on until I hit the jackpot too. Unfortunately, I hit it during the very last week of the year. There were no more papers anyway. Talk about lousy timing.

I don't know if that sort of incentive teaching goes on anymore. But I do know that in recent years dozens of corporations have become so frustrated at the shortcomings of the people joining their work forces that they've been driven to provide remedial education on their own. It's estimated that as many as a hundred companies—including Eastman Kodak, AT&T, and Polaroid—now offer courses that can lead to accredited degrees. American companies cough up around $40 billion a year to teach 8 million workers. Our colleges and universities, meanwhile, spend only $60 billion to educate roughly the same number.

Some corporations have taken their in-house efforts to the point of bestowing actual degrees on graduates: You can now earn the Arthur D. Little M.S. in Management or the Rand Graduate Institute Ph.D. in Policy Analysis. But I don't feel American business ought to usurp the role of our schools. I don't want to see Chrysler teaching fractions and world geography. Before you know it we'd have to offer gym, and field a football team.

What business can do is support our schools and make sure our teachers are tuned in to the real world.

On November 6, 1986, I got two letters in the mail from the presidents of two universities—Lehigh and M.I.T. When I read them, I thought to myself that maybe they'll be historic documents someday, because both presidents had come to an identical conclusion. It had dawned on them that their schools had taught many of this country's professors. M.I.T. alone turns out 11 percent of all the engineering professors in the United States. It's an amazing number. Now these schools suddenly realized that those professors have been teaching students the wrong stuff.

Basically, the message they've given our kids is to go out into the world and make a quick buck for themselves. Nobody was stupid enough to advise a kid to go out and get his hands dirty by making something and helping this country become productive. Only uneducated fools would do that. If you've got the business smarts, they say, go to Wall Street and make a killing; if you're a technical whiz, they send you to a Southern California defense contractor to work on black boxes for the military—neither of which has anything to do with productivity.

They tell me we've got twenty-three thousand of the finest minds in America at Hughes Aircraft working on Star Wars and other classified projects. The Japanese, on the other hand, have an equivalent twenty-three thousand minds working on sharper TV pictures, graphite golf clubs, and electronic instrument panels for cars. They have nothing else to do, since we won't even allow them to work on any sophisticated military gizmos. That's a brain drain that boggles my mind. Our best and brightest put the results of their work into missile silos for storage. The Japanese bring good things to life for the needs and enjoyment of consumers.

Just try to recruit a new graduate from the Harvard Business School or M.I.T. to come work in a car company. He typically tells us: "Go work for Chrysler as a financial analyst? You must be out of your mind. I'm going to Wall Street." And we're not just offering him crumbs. We pay him a cool $50,000 to start.

It's the same story with engineering graduates. "What challenge can there be in trying to build a better car?" they ask. "That's a dying industry. The action is in laser beams."

The real irony to me is that the guys in California getting paid by the Defense Department are the very guys who buy all the Japanese cars on the theory that American cars are not as technically advanced.

In counseling our students, a lot of teachers and advisers have followed the flow of the money. If that approach undercuts the country and causes the industrial base to go to pot, they don't really care. I seriously doubt they've ever given it a second's thought. Their spiel is: "Kid, I've taught you all I know. I want you to be a success. Your becoming an investment banker (or a rocket scientist) will look awfully good in the alumni record." After all, how do we judge success in this country? First and foremost, money.

To give you a feel for how things have changed, ten years ago Lehigh asked its graduating seniors whom they wanted for a commencement speaker. Their choices were Walter Cronkite, Bill Cosby, and me, in that order. Cronkite and Cosby turned them down, and I accepted. (When I showed up, I said, "Hey, kids, I hear I was your third choice. So what you're going to get is a third-rate speech.") Last year, when Lehigh asked the seniors whom they wanted, the picks were Clint Eastwood, Steven Spielberg, and Donald Trump: the movies, make-believe, and easy money.

The schools are having the *mea culpa* of their lives. They're finally coming to the realization that they've created a monster.

They've figured out that we have our priorities screwed up: We have to train people to do every job, not just the ones in the financial district. The manufacturing process has been sadly neglected in this country, and now's the time to really throw the coals to it.

And so, both M.I.T. and Lehigh concluded that they needed to develop a new curriculum to tackle our competitiveness problem out there in the world marketplace. Lehigh's idea was of particular interest to me, since they claimed that I inspired it. Peter Likins, the president, wrote: "Wouldn't this be a wonderful place for you to spend the rest of your life? You've become synonymous with making America productive again. Why not give it a shot here? We will not only start a whole new curriculum, we'll start a whole new college."

Yup, Lehigh has created the Iacocca Institute.

By now everybody's got a different cliché to sum up our problems, but nobody really knows exactly what those problems are. In this new school, Lehigh wants to zero in on what it is that makes a country an industrial power. What does preserving our industrial base mean? Which industries should be kept and which ones discarded? What should our companies look like five years from now? Ten years? By setting up cooperative projects on the campus between companies, academia, and government, and by crystallizing the question of productivity into actual courses, Lehigh hopes to come up with some answers.

The setting of the institute is the irony of the century. Lehigh's campus sits on a mountaintop. Twenty years ago, at the very top of the mountain, the venerable Bethlehem Steel Corporation built a research center in order to insure that the company would stay in the forefront of steelmaking. Apparently the center didn't do the job. Today Bethlehem is struggling, part of the saga of our declining steel industry.

As Bethlehem's business crumbled, one of the measures it had to take to stay alive was to shuck off the R & D center. That's always the first thing you get rid of in a crunch. It helps companies short-term but sure knocks the hell out of your future.

The state of Pennsylvania and Lehigh anted up $20 million to buy the center—and that's where my institute is today. Of course

I also agreed to help raise $40 million to do all those nice things they want to do. It's amazing how a mere $40 million will get your name on a building—or even a college.

Maybe the history of the institute should be the first big case study. Some students may be curious as to why they're doing homework in an abandoned R & D center of the Bethlehem Steel Company. There's a lesson in there somewhere.

If the institute succeeds, I'd like to see the idea spread to twenty universities so that we'll all be able to plug away at the question of what it takes to be competitive. If everyone swaps notes and learns from one another, the result may be a breakthrough, like the way it works with a disease. Because, make no mistake about it, the disease of our industrial decline will be fatal without a cure—and soon.

XIV
IN SEARCH OF QUALITY

We've all had the experience. A coffeepot that takes six hours to heat the coffee. A microwave oven that explodes when you put the baked potato in it. A TV set that gets only one station—and it's the Lower Slobbovian one. A new car whose wheels fall off before the customer gets home. (Not that I would know anything about that!)

What's the common denominator? Lousy quality.

Every day when I go to work, one of the things that drives me is the challenge of trying to squeeze more quality into an automobile. I've told my managers a hundred times: The only job security anybody has in this company comes from quality, productivity, and satisfied customers. Those are the arrows in our quiver. Without them, you don't put meat on the table.

The plain truth is that this country has really let quality slip since our heyday of the 1950s and 1960s, when "Made in the USA" meant the absolute best in the world. And the slippage has gone way beyond the coffeepot and the TV.

There's been a huge shift in our national attitude. I'm convinced that, deep down, my generation has something of a guilt complex. Even though we've accomplished many wonderful things—we managed to wipe out a few diseases and put a man on the moon—there seems to be a feeling gnawing inside of us that we've screwed up and have to settle for second-best, that everything has gotten worse. The pipe under the sink is leaking, and the plumber shows up three days late and needs hip boots to wade past the floating pots. The electrician leaves $200 richer, and when you turn on the bedroom light the furnace starts up. It's no wonder that housewives are becoming crack electricians. They're frustrated and they can't take it anymore. And to top it off, the kids are still yelling and screaming and the dog hasn't been fed, so they're about to go bananas.

There's no longer any pride, even by professionals, in getting something done. Salesclerks are either too busy or too snobby to help. Or they're so badly trained that they could spend all day with you and still not find the Hot Lips lipstick. If you do get a cheery salesperson, you think the person's on drugs. Airplanes have fewer flight attendants than they used to, and they all seem grumpier. By the time you're served your meal, the seat-belt light is on and you've probably lost your appetite. And the meals? Better never to eat and fly at the same time.

Years ago, retailers really doted on customers. Record stores used to have record players available so that you could listen to the records before you bought them. Now you're lucky if you can find the record you want. If you do, you're really lucky if it isn't warped.

The movie *Back to the Future*, in which Michael J. Fox journeys back in time, has a funny but telling scene. Fox wanders past a 1950s gas station and is shocked at the sight of four smiling attendants converging on a car to fill up the tank, clean the windshield, check the oil, and even polish the chrome. Today you pump your own gas, and the attendant acts as if he's done you a favor when you hand over the money.

Look at the quality problem NASA had. Who would ever think we would lose a space shuttle because of a lousy O-ring? We don't even use O-rings on cars anymore. That whole project went haywire over one monumental mistake in design.

It's distressing to think that we've regressed to this point, because America virtually invented quality products and services. AT&T,

IBM, Federal Express—they're synonymous with dependability and service. (Although I have to admit, when we deregulated AT&T we did manage to screw up the finest phone system the world had ever known.)

What went wrong? Well, I can't blame everything on Vietnam and Watergate, but I think that the one-two punch of those two tragedies helped trigger this decline in quality. Vietnam was an immoral war. It was followed in quick succession by an immoral presidency. Seeing cheats and liars in positions of leadership caused people to say: "Forget it. Nobody gives a damn anymore."

Wait a minute, you might say. Are you trying to tell me that Vietnam and Watergate had something to do with the fact that I can't get my TV fixed? Well, it's hard to draw that conclusion from something so broad, but there is a connection. The country just plain lost its pride. The feeling was: "There was our proud army that never lost a war, and now look what happened to it. That was my own government, and look at these guys—they're a bunch of crooks." Such depressing events had to affect the psyche of the country.

On top of that, the high inflation of the 1970s forced a lot of businesses to cut quality and service in order to keep prices from skyrocketing. Then deregulation brought about still more cutbacks. To make matters worse, service workers became harder to hire because of labor shortages in many parts of the country. Then computers, automatic tellers, and self-service gas stations did away with a lot of workers, and the human touch became a very scarce commodity.

Many companies kidded themselves that their profits would shoot up if they cut corners in quality. Well, plenty of industries, including the American auto industry, found themselves not gaining but losing market share because of shoddy products. If there's any doubt that lack of quality costs American industry a ton of money, get this statistic: As many as one out of four factory workers produces nothing at all. They spend their entire workday fixing the mistakes of other workers.

When your company is shelling out a quarter of its operating budget to clean up the foul-ups of others, the message sinks in fast: The only way to compete is by making better products.

* * *

Quality, of course, is a fuzzy concept. It's almost like a work of modern art—difficult to define because it means radically different things to different people. Reduced to its simplest terms, though, quality means products that work well and last long. Quality means a toaster that browns both sides of the toast, and a washer that gets rid of ring around the collar.

For a lot of people, quality also means convenience. Witness the growth of mail-order catalogues and the rise of home shopping through your TV. Quality means an airline that gets you to your destination before the seasons change. And it means the hope that someday you will again meet up with your luggage.

It also means service with a smile if something goes wrong with a product. If anything mechanical breaks, people want the manufacturer to stand behind it. That's why Americans have become so big on warranties. The consumer knows that while the company is correcting its quality problems, at least he's covered for the expense. That's what we in industry call quality of service after the sale.

I'm supersensitive to concerns about quality, because all my life I've been in a business that has been much maligned (as well as much congratulated) for the quality of the products it builds. Our customers work hard for a living and we take a big chunk of their disposable income in exchange for something we tell them they will be happy with for years. If it doesn't live up to their expectations, boy, do we hear about it—and I mean loud and clear.

In the real estate business they say the only three things that matter are location, location, and location. In our business it's quality, quality, and quality.

We know who decides whether we're the best or not—and that's the customer. He or she is the referee in all of this. Nobody else counts. *Consumer Reports* may have its opinions on who's best. We have our own opinions. But the only guy with a vote is the customer—because he's the one who lays out the money. You listen to him carefully or you die.

The customer may not appreciate what goes into quality, but he sure understands what comes out. He may not know a transaxle from a turbocharger, but he knows what *fit* and *finish* mean. He knows what a quality feel is when he slams the door. And he sure knows what reliability is. It's simple: "I hope to God the car starts in the morning."

And one thing you can bet on. The customer is never neutral on quality. It's either a great car or it's a lemon. There's not much in-between.

After all we went through in the past at Chrysler, I'd like to run out and buy an insurance policy to cover our future. But guess what? Nobody's selling. You can insure yourself against the guy who sets fire to the plant. You can insure yourself against the guy who sues you. If you pay enough, you can even insure yourself against an act of God. The only guy nobody will insure you against is the customer. That policy we have to write ourselves.

How do you write it? First of all, you have to understand that quality is an attitude. It doesn't have a beginning or a middle. And it better not have an end. The quality of a product, and of the process in arriving at that product, has to go on and on to become part of every employee's mind-set.

It all depends on people. Sure, we need those expensive robots and space-age lasers, but they're just tools. Quality isn't something you can buy; it's something you must attain—through people. The quality improvement process is just ink on paper until workers breathe some life into the process.

This sounds corny as hell, but everybody in an organization has to believe that their very livelihood is based on the quality of the product they deliver. They must truly think that quality is the only thing that buys them groceries and pays their mortgage and puts their kids through college. And they have to go after it with all the fervor of converts to a new religion—and you know how they can be. (I never saw fervor to equal that of the guy who's given up booze or who's all of a sudden found God. It's murder on the rest of us!)

In Japan the commitment to quality is so ingrained it's almost like personal hygiene. And that's got to be our commitment too: to make that goal so much a part of our thinking that we don't have to think about it anymore.

To do that, the boss must be absolutely intolerant of any shortcuts. If he gives lip service to quality—meaning he knows something's wrong but he keeps building it anyway—then the word spreads like wildfire. Workers at every level will say: "Well, if the boss doesn't give a damn, why should I?"

This may seem simple enough to implement, but any good manager will tell you that in fact it's incredibly difficult to instill a belief in quality in a work force. You don't just call the troops into the auditorium, flick out the lights, and give them a slide show.

Maybe the hardest part about making quality the base of our whole culture is the fact that we have to do it by trial and error, often in the dark. We also have to do it with moving targets—and anything moving in the dark has always worried me.

Quality, after all, is affected by something as basic as a person's sense of values. Maybe a guy working on the line doesn't even know his job. Or maybe he's not trained well enough. It could be that he got bombed over the weekend and came in Monday just trying to get through the day.

Remember the old adage: Don't buy a car on Monday or Friday because that's when the worst cars come off the assembly line? Well, there was a grain of truth to it in many of our big-city plants. But it's not that way anymore.

If a person's going to do a good job, he's got to like coming to work. He's got to say to himself: *I'm going to help produce something great today*, and he's got to say that every day.

I like to go to work in the morning. I've been doing it for forty-two years now, and I'm still as enthusiastic as I was the first day. Why? Was I born that way? Was it my teachers? Was I a screwball? I suppose it was a lot of things.

Unless a person develops that enthusiasm, he'll never understand his part in making a better product. And if he doesn't, everyone's going to get into hot water.

Every day in America, 242 million people wake up, and if everyone would say when he gets up that he's going to do some classy, quality thing today that he didn't do yesterday, we'd be worldbeaters. Unfortunately, most people swing out of bed, yawn, and figure: *Oh, hell, I've got to make it through another day of drudgery.* Their attitude is that they're going to do what they're told and not one thing more. Now, how can you ever improve anything that way?

Through all my years in the auto business I've been hearing manufacturing guys complain that the stupid designers made the car impossible to build. It's got eight pieces, they say, when you could just as easily design it with two. And the designers lament, "Look, we gave them a perfectly good design. They're just too dumb to know how to put it together. It's their problem."

And all this within the same company: the ongoing belief that any problem is somebody else's fault. The first thing you've got to do, from the top, is to make people understand that the quality of the finished product is everybody's job. Every single person in the plant has a hand in it, down to the last guy who fills the car with gas and washes it. If gasoline runs down the side while he's filling it up, it'll ruin the paint job that somebody else carefully applied. The product that goes out the door has every worker's fingerprints on it, and I don't mean greasy ones.

Let's face it, quality improvement isn't headline material. Reporters descend like locusts every ten days to get our sales figures, but they never call up to say: "Hey, how is your quality coming along the last ten days?"

So this is a job we have to do with no bands, no cheerleaders, and no glory. Just months and years of first-class discipline and hard work. The fans in the stands in this case have to be the workers; they've got to refuse to accept anything second-rate.

Think about it. Just twenty years ago "Made in Japan" meant cheap and shoddy. Even the fireworks were lousy. Now "Made in Japan" means quality and good value. I picked up *The New York Times* one day and saw a public opinion poll that really scared me. It said Americans believe that the Japanese are better workers. One guy said, "We're very bad about waiting for twelve o'clock and five o'clock [that's lunch and quitting time] instead of doing our jobs." Another person declared, "Many American workers don't have the dedication to go the extra inch if it's not in the job description." Or: "I guess the Japanese are just more determined to get ahead." By the way, if I hear one more guy say to me, "That's not my job," I think I'll go mad.

These were Americans talking about Americans. Whatever happened to pride in this country? There are some things you just don't say out loud—even if you believe them. Where are we going if we really believe these things about ourselves? How are we ever going to compete if right from the start we're saying "The other guy is better than I am?"

The worst part of all this is that nobody seems too alarmed about it. Can you imagine the ruckus in Washington if they took a poll and Americans said the Russians were better soldiers?

The American bias against American goods has to go. The Japanese will buy higher-priced or inferior products if they're Japanese-made. We're just the opposite. We believe that anything foreign must be better. Someone once wrote me a letter saying he couldn't understand why so many Texas ranchers drove Toyota pickups when the Japanese had such strict quotas on Texas beef.

It's one thing for an American to rave about French wine or Swiss chocolate. But we invented mass production in this country. It built the middle class. It gave us a standard of living unmatched in the world. And now Americans are saying that Americans can't hack it anymore.

I don't want to stretch this too far, but the American consumer seems to put what he considers quality ahead of the flag. And maybe that ought to tell us just how important quality really is to him.

The good news is that we're starting to see a lot of efforts being made around the country to improve. There are now dozens of quality gurus who, for a handsome fee (I know, because I've paid some of them), offer companies advice on how to wring more quality out of their products and services.

For my money, nobody talks quality better than Phil Crosby, a well-respected consultant who even has a quality college down in Florida. We thought enough of it that we established our own Chrysler Quality Institute in Michigan, modeled after his operation. Our company's put about twenty thousand of our people through it—they're going back to school at a rate of nearly two thousand a month—and I must admit they do return with QUALITY stamped on their foreheads.

We can't ever let up, though, because our competitors are also pushing very hard. They aren't asleep at the wheel. Their drive for quality is as real as ours, and they've been pretty successful too. In 1980, when Detroit saw how far ahead Japanese cars were in quality, the Big Three kicked off crash efforts to cut the gap. Quality studies now bear out that a car made in America is right up there with a car made in Japan. Not quite as good, but almost. Yet because of past sins, the public's perception of our quality lags far behind the real thing. If our quality weren't up to snuff, Chrysler sure wouldn't be offering a seven-year, seventy-thousand-mile warranty, the best in the business. That's putting your money where your mouth is.

Big improvements are showing up in such basic appliances as refrigerators and stoves. Last year GE agreed to take back any major appliance for a full refund within ninety days if the customer didn't like what he or she got. Whirlpool went them one better and promised to replace any of its major appliances free of charge within a year if the customer wasn't completely happy.

In fact, everybody is pushing quality so hard that, if this keeps up, maybe one day the ideal we call "defect free" will become the norm. I can assure you that whichever company first sets that standard will really be in the driver's seat.

Even cities and municipalities are jumping into the quality act. In Miami, all of the city's five thousand cabdrivers are required to take a three-hour course in courtesy called "Miami Nice." The training seems to do the trick. Customer complaints have plummeted by 80 percent since the course started. (Whether the approach could ever work with New York cabbies is something else again.)

One of the nice things about quality is that when most red-blooded Americans see it, they like to brag about it—they're so grateful. I'll never forget the dentist in Detroit who drove his New Yorker over to our Jefferson Avenue plant and parked it by a group of line workers who were outside on a break. He walked up to them and said, "Did you people build this car?" They didn't know what to expect—maybe a tire iron across the skull or a can of Mace in the face—but they owned up to it. Then the dentist said, "I just want you to know that it's the best car I've ever owned in my life and you people oughta be proud for having built it." Then he shook hands all around and drove away. A kook? No. Just a grateful American customer—and morale-booster.

Boiled down to its essentials, quality is so important because it ultimately determines how competitive we are.

We've got to compete in the global economy we did so much to create. And we aren't king of the hill anymore. We can't dictate the rules. We don't have a lock on technology or new ideas. Now we're just one of the players. And so we've got to roll up our sleeves just as our fathers and our grandfathers did.

Competitiveness, after all, begins with the ability to produce goods better than anybody else in the world at a price that's attractive to consumers. Everybody wants the best there is, and they want it at the lowest possible price. That's called value. Believe me, if you're able to give people honest value, you'll always be competitive.

In 1986 we showed everybody—including ourselves—what working together can do. We did it with those little (twelve-year-old models) Dodge Omni and Plymouth Horizon "America" cars. Yes, we stunned everybody when we put them on sale for $5,499, a sticker lower than on any American car—or any Japanese one, for that matter. But we hadn't lost our minds, and we weren't kicking off a fire sale. We intended to make money at $5,499, and we did.

We did it because we worked together. At the corporate end we cut the price on the Horizon by $710 and then threw in $689 worth of equipment so that no one could say it was an "el strippo." The union agreed to more productive work rules. The State of Illinois gave us some training money. Our suppliers cut prices by about 10 percent. And our dealers came to the party by agreeing to smaller margins.

We all worked together, and look at the results: Since the America cars went on sale, our share of the subcompact market has almost doubled. That speaks for itself, and it also proves that Americans will "Buy American" if you give them a quality product at the right price.

*   *   *

There's another side to quality that nobody ever wants to talk about. But I'm a dollars and cents man, so I'm going to talk about it. Quality is not free. I think I know quality when I see it, and it has a hefty price tag.

Some people—the consumer activists in particular—don't think American business is going fast enough in the quality department. But they forget one thing: There's a price for speed.

What's more, you have to stop and think about how much quality you really need. Some women spend $5,000 for a dress, which is nothing more than a simple piece of cloth. What if they spill something on it? I've noticed that Countess Mara ties now sell for as much as $200. They're not for me. I'm constantly spilling gravy or something on my tie and end up tossing it in the garbage.

So how much quality are you willing to pay for? Well, can you ever build the absolutely perfect car? I guess you could, but the car might cost you $100,000 instead of $10,000. I'm not too sure that even Rolls-Royce has the perfect car at over $100,000.

If you really wanted it, I'm sure a suit manufacturer could give you so much quality in a suit that it would last you twenty years. It would be warm, it would wear like iron, and it would never drop a stitch. Oh, by the way, it would cost $8,000. Wouldn't forty suits at $200 be a better buy—and a helluva lot less boring than just one?

Unfortunately, a lot of people in this country have the attitude: "I want quality, but I don't want to pay for it."

Well, guess what? You can't have it both ways. And so if America is really serious about competing in the world, then we're going to have to be willing to pay some stiff costs.

Everything that needs to be done—from cutting our budget deficit to improving our schools—is going to cost money and it's going to mean sacrifices. If we as a nation keep ducking the costs, then all the talk about "competitiveness" we're hearing today is just a lot of hot air.

But while we address those costs, there's a second question we have to ask ourselves: Just how competitive do we want to get?

Japan is always held up as the model of competitiveness, and there's no doubt that the Japanese have paid some heavy costs to earn that distinction. Their banks are full of money but their tightly controlled economy with its "export but don't import" philosophy means that a housewife pays $50 a pound for prime beef and six times the world price for a bowl of rice.

The newly industrialized countries like Korea and Taiwan are setting the world on fire with their high-tech exports, but the factory workers who make them earn $2 or $3 an hour, and that sure doesn't buy them a very high-tech life style.

It's more than our wage rates that push up the price of American products and put them at a competitive disadvantage with those from the Far East. We also charge the consumer some "social costs" that the imports don't have to pass on.

We're one of the few nations in the world whose people can sit around and debate the quality of life and what we want out of it—materially, that is. Some nations can't spend even a few seconds on this subject. They're so worried about the basics, like eating every day and having a place to sleep at night, that they have to forgo the luxury of quality products and service with a smile.

That's why Brazil has been making such a big pitch for our paper mills to invest there. With all the environmental rules we have in the U.S., paper mills are practically banned. So Brazil is saying: "Come down here. Give us your paper mills. Pollute our streams. Dirty up our country. We can't worry about that right now. We need jobs. We need to eat. Then we'll worry about the qualitative things that you folks worry about up there."

We don't do it that way in this country. For example, the price of a new Chrysler car includes more than wages, and more than steel, rubber, and glass. Somewhere hidden in there is the price of our standard of living.

It doesn't show up on the sticker in the window, but part of that price goes for employee health-care costs. In the United States, employers pay most of the health-care bill. In other countries, that's mainly a governmental responsibility. I like our system better, but it does put a burden on the competitiveness of our products. Right now, our health costs are running about $600 a car.

We've got programs to hold those costs down, but they're still high and they always will be, because providing workers with top-quality health care is one of the costs of doing business in this country. But I'd like to point out that those free braces for all the kids' teeth and prescription sunglasses don't come for nothing. You can keep them only if you're willing to offset the costs by doing your job better.

Every car also includes employees' pension costs. In many other countries, that's a government responsibility. And in the most "competitive" (translate that as low-wage) countries in the world, there are no pensions. An American worker's compensation covers not only his paycheck but his retirement as well. And the cost has to be passed on to the consumer. That's another $275 a car.

Here's more: We've decided that clean air and water are the employer's responsibility. The Environmental Protection Agency sets the standards for our factories and polices them, but almost all the costs are folded into the price the consumer pays for American products. Sure, we can argue over the details and how you implement our environmental laws, but nobody wants to foul our air and water just to undercut foreign prices. Add another $99 to the price of a car.

Sweatshops aren't legal here anymore. Companies spend billions every year to make their facilities more productive and at the same time safer. It's impossible to break down exactly what health and safety standards add to the cost of a car, but the dollars are substantial—probably in the $50-per-car range.

I must admit, I have to wonder a little about what we're getting for some of those costs. Last year Chrysler had a well-publicized run-in with the Occupational Safety and Health Administration, or OSHA. We wound up paying a $1.5 million fine for infractions at just one plant, although we admitted no guilt.

As far as injuries on the job are concerned, the auto industry happens to rank way ahead of most others on safety. And within the Big Three, Chrysler is just as good as the others. Yet, despite our record, OSHA decided to take a close look at our Newark plant. It's an old plant, which admittedly is no excuse, but no plant in this country can possibly pass perfect muster under the endless OSHA regulations. Remember, this is the organization that even has rules on toilet-seat and urinal heights. When their people come in and give you the white glove inspection, they're going to find violations in any plant in the United States, even if it was built yesterday.

Well, they found some at Newark. Case in point: OSHA discovered some leaks in the roof. Fine. We told the agency we'd put cans under them until next year, when we could afford to redo the whole roof. "Oh, no," OSHA said. "That's a safety violation of the first order and it can't wait. Correct it now."

I'm not an advocate of leaky roofs, but they really put a short leash on us.

Another example: We don't have many robots in that plant, but we have some. The rules say that you must have a steel fence around them so that nobody can inadvertently walk up to the robot and get eaten up by mistake. In an old plant, we're not going to spend a great deal of money to put up steel guards, so we put up rope rails. They served the purpose fine. No one's ever gotten hurt. Forget it, said OSHA: violation, fine—correct it.

OSHA did find one thing I can't condone for a second. Newark is one of only two Chrysler plants left that use solder. We've spent millions of dollars to get solder out of the plants, and the final two will change over this year. Anyone working with solder is exposed to its arsenic and lead content, which can be harmful to the bloodstream. OSHA's code says that guys working with solder must get their blood checked on a regular basis.

Well, we send our blood samples to labs in Detroit, and one batch took a couple of weeks to check because the blood machine was busted. Two of our workers had a high reading. They weren't told in time and got dizzy on the job. I wish we had done a better job with that one.

But OSHA latched onto that screw-up and said it was typical of hundreds of violations it had turned up. It didn't bother to mention that virtually all the other violations were nitpicking.

To make sure we didn't get the same treatment at our other facilities, we formed SWAT teams and made believe all of them worked for OSHA. Those teams took the rule book and in four weeks went through all the plants at a cost of millions of dollars, correcting every violation on the spot. You can bet the plants are now up to snuff.

Even after all that, however, I'm sure OSHA inspectors could get down on their hands and knees with microscopes and still find a violation or two.

I'm certainly not quibbling over any violation that endangers someone's health or safety. If we've been careless, we should correct the exceptions and pay up like anyone else. But not all violations are created equal. And we really have to get off that kick that equates any hazard with setting your mother on fire. Neither I nor anyone else wants workers to get hurt or get old before their time. But you can take the idea of protection entirely too far.

Last year, the Big Three did something unusual. For the first time ever, they signed a joint letter vehemently protesting a decision by the EPA that would needlessly add to our costs. In a handful of areas in the country, such as downtown Los Angeles and Detroit, the EPA had declared that there's a smog condition created by evaporative emissions. When you pump your gas at the service station on the corner, that smell you recognize is evaporative emissions, which go into the air and affect the ozone in that general area.

Years ago the auto industry installed a small canister on all cars to try to collect the fumes of the on-board gasoline. The EPA now wants us to triple the size of that canister. They maintain that the cost to do that is $19 a car. We and other manufacturers believe it's actually $60 to $120. What's more, it's going to take twelve to fifteen years to get rid of the problem by increasing the canister size of new cars, because it's going to be that long before we've gotten rid of all the old cars that don't have them.

We argued that there should be a trap on the pump nozzle at the gas stations. Forget it, the EPA said, claiming that would cost three times as much. So what? Then the problem would be licked in one year, not fifteen. And it would have to be done in only a few regional areas, anyway, so the total costs to the nation would be much less. We're still fighting the case.

The point is, life isn't perfect, and it never will be. As a society we have to be vigilant against environmental atrocities. And yet we're going to have some fumes in the air and we're going to have stuff dropping into streams. Some people believe that if a little regulation is good, a lot is much better. Before you know it, regulation takes on a life of its own and strangles you.

Besides these and other social costs, we also pay a premium to maintain some rights that other countries don't care too much about. One of them, as I've already mentioned, is the right to sue each other at the drop of a hat. If Americans want the right to sue the doctor, the drug company, or the auto maker, then they have to pay for that right in higher prices.

At Chrysler we have to plug those liability costs into the price of each car. In Japan, firms don't have that problem. They just don't sue each other (although when they sell their products here, we sometimes sue them!).

The moral of this story is: America must get more competitive. But how competitive do you want to get—and how much do you want to pay for?

Do we want to compete against Japan by adopting its life style, or against Korea by cutting our wages to two bucks an hour and tossing out our environmental and safety laws? Are we willing to trade a big part of our standard of living to compete?

I don't think so. If competitiveness is reduced to simply "who can produce it the cheapest," then the biggest winner will be the guy who's willing to be the biggest loser.

OUT FROM UNDER

XV
TWENTY AND EIGHT

Life is full of all sorts of cycles. Some of them are predictable—night follows day; fall follows summer; tides follow the moon. Those are the ones God takes care of. Then there are the ones people take care of: business cycles, energy cycles, automotive cycles. Those we manage to screw up but good.

It would be great if we could repeal some of the man-made cycles or somehow shoo them away, but the best we can do is take a stab at moderating them a little so we don't get banged around too much. That's one of the things we pay the people running our government to do—to write policies that smooth out the most extreme ups and downs. In some cycles you can't just let nature take its course.

As the head of Chrysler it's my job to ask: "When things are going good (or bad), what happens next?" Well, as a citizen it's your job to ask what happens next to your country.

I try to answer that question by looking for patterns. As far back as I can remember, I've always been a strong believer in the importance of cycles. You'd better try to understand them, because all of your timing, and often your luck, is tied up in them. I've even formulated a little theory of my own about them, which I call "Twenty and Eight."

In a nutshell, the theory is that the country tends to stagger through twenty years of havoc and activism and then to settle into eight years of relative calm. After catching its breath for eight years, it goes through the wringer again for another twenty years. And so on and so on.

The theory shouldn't be taken too literally. It doesn't mean that during the eight years of calm there's not one problem, not even a traffic jam somewhere. In fact, big trouble seems to start kicking up toward the end of each of those stretches. And naturally, there are going to be some high points during the twenty-year downers—it's not nonstop misery.

I guess the best way for me to illustrate the theory is by using the framework of my own lifetime.

I was born in 1924, when Calvin Coolidge was President and the country was heading into one of its periods of calm. When I started noodling with this theory, I had to go back and read a little bit more about those days, because I didn't have a close eye on Coolidge in 1924. After all, I was lying in my crib. My parents didn't interrupt me in the middle of my two o'clock feeding to tell me about how Cal was doing in Washington.

When Coolidge took office, after the trying years of the First World War and the early 1920s, the country was long overdue for a good shot of tranquillity. "Silent Cal," of course, didn't want to be President; Warren Harding died and Cal simply got stuck with the job.

Coolidge ran the country according to the same naïvely optimistic beliefs that Ronald Reagan holds. He once said, "If you see ten troubles coming down the road, you can be sure that nine will run into the ditch before they reach you."

Until recently, I didn't understand why, when Reagan won the election, he hung a picture of Calvin Coolidge in the White House. That was pretty weird. Most Presidents, if they're Republican, nail up a picture of Dwight Eisenhower. But Coolidge? The guy known as one of the great thumb twiddlers of all time! Yet Coolidge was

Reagan's mentor. Reagan admired him more than any other President, and he imitated him perfectly. After Reagan, Coolidge surely has to go down as the greatest hands-off President in history.

Cal was a man of few words—a sentence was a lot of talking for him. There's a classic story about Channing Cox, who succeeded Cal as governor of Massachusetts. Cox dropped in on Coolidge and asked Cal how he managed to see all his visitors every day and still leave his office at five, while Cox could rarely get away before nine at night. Coolidge replied: "*You* talk back."

Another time, a dinner partner told old Cal she had made a bet that she could get him to say more than two words before the dinner was over. He said, "You lose."

When Cal said anything at all, it was to urge everyone to do his own thing. As far as he was concerned, the less government interference the better. And so we had the Roaring Twenties, the flappers, and all that jazz. After the recession of 1920–1921, people needed someone like Cal to offer them a respite. And he filled the bill perfectly.

Unfortunately, that hands-off approach also planted the seeds for the great crash of 1929. As the country drifted into the late 1920s, the wheels started to come off one by one. There were abuses galore. A raging stock market. Scandals in the government. In 1929, with the walls cracking around him, Cal left office.

Along came Herbert Hoover to try to patch things up. He was a mining engineer—in fact, the last scientific President until Jimmy Carter—and people felt that those precarious times demanded a businessman who was a logical thinker. Hoover, of course, went down in the history books as a very weak President, which was probably a bum rap. The truth is, things were reeling out of control and poor Hoover batted cleanup for Coolidge. He was the fall guy.

Hoover promised to deliver what he called "The New Day," an era of great economic and social prosperity during a period of developing science (today we'd call it high-tech). He attempted reforms in the antitrust laws, the stock market, banking, and in the regulation of electric power. Unfortunately, he brought less political experience to the job than any President since Ulysses S. Grant, and so everything he tried fizzled. (With Grant, everything just gin-fizzled.) As Hoover once wrote: "I had little taste for forcing congressional action or engaging in battles of criticism." He must have forgotten about the balance of power that the Constitution had in mind.

Although he did his best, it was just too much to expect him to make the downward slide stop on a dime. As always, life was far too complicated for fast fixes.

The state of the economy worsened, and then, in 1929, the market crashed and all hell broke loose. People started selling apples on the street corners and jumping out of windows on Wall Street. Unemployment soared to 25 percent, then 28. Tens of millions of people found themselves in dire poverty. The whole world dived into a Depression and no one seemed to know what had caused it. There was a lot of finger-pointing, and most of the fingers pointed at Hoover. In many people's eyes, he was a bum. Throughout the nation, Republicanism became a dirty word.

Now it was '30, '31, '32, and the country was really bleeding. Pretty much everyone was screaming to get rid of Hoover. A number of taunting remarks about him became popular. One of the favorites went: "He said that prosperity was around the corner, but he forgot to tell us which corner."

With Hoover on the ropes, the country was looking high and low for a white knight. It found him in FDR.

I have my first clear memory of a President from around this time. In the ethnic neighborhood in Allentown where I lived, many families were pretty hard up, and so automatically they were Democrats and wanted a change. FDR's election pitches roused a lot of interest—and emotion.

The first time I remember hearing the names Hoover and Roosevelt was one day at school. (Until then, Hoover, to me, was my mom's vacuum cleaner.) Three buddies and I were standing on the schoolhouse steps in Allentown. I even remember that I had on a pair of corduroy knickers. My friends and I were chanting a little ditty just for fun. It went, "Two, four, six, eight, who do we hate? Hoover. Hoover." Followed by, "Two, four, six, eight, who do we appreciate? Roosevelt. Roosevelt."

When my friends started singing it, I joined in without even knowing why. Since we were all in the third grade, I couldn't have learned it in political science class. But that silly experience marked my first consciousness of a President and a political system in this country. I didn't even know what a two-party system was until then.

Like others in my neighborhood, my father was quite an activist and an FDR fan. Even though he didn't have an abundance of education, he was always writing letters to the editor, always writing to the President of the United States, always giving his opinion. (I guess I come by it honestly.) He told me that Roosevelt was the man we needed in Washington. I was a bit too young to make any sort of reasoned judgment of my own—I mean, what interested me was eating more candy bars, and nobody seemed to be running on a more-candy-bars-for-kids ticket—and so I just picked up what I heard at home.

FDR wasn't a poor guy—he had been born into wealth—but he came in and said right off the bat that we've got to change things. FDR's election marked the beginning of the New Deal and the era of very active, hands-on government—like it or not.

With the roof falling in on his head, FDR had no choice but to take charge. After all, in 1932 the whole world was sunk in a Depression. I'm not a historian and I don't want to try to analyze the whys and hows too deeply. But inflation took hold in Nazi Germany during the 1930s, and when economic things go to pot, political vacuums are created and in come the bad guys. In the '30s, Europe was in that kind of turmoil; it was changing radically and picked up more than its share of rotten apples—Adolf Hitler, to name just one.

Roosevelt set out to put some order into the chaos of the financial markets. As he vowed in his 1932 campaign: "It is common sense to take a method and try it. If it fails, admit it frankly and try another. But above all, try something."

In the Hundred Days—those tumultuous weeks from March to June 1933—Roosevelt tried plenty. He sent fifteen messages to Congress and pushed fifteen major laws to enactment.

If he hadn't, this country would really have been blown out of the water. There had been widespread abuses of buying stocks on nothing down; in fact, speculators were playing paper games just as they are today. So FDR gave birth to the Securities and Exchange Commission. He told us that a society should take care of its old people. Enter Social Security. He brought about the Tennessee Valley Authority and the Civilian Conservation Corps. In those Hundred Days he introduced a lot of tough measures that proved essential to keeping the country on its feet—instead of its last legs.

We were forced to drink so much castor oil in the first couple of years under Roosevelt that even my father, the great Roosevelt-phile, nearly turned Republican. The National Recovery Act was what really bugged him. At that time he owned a hot-dog restaurant called the Orpheum Wiener House. The act required him to pay a minimum wage to a waitress and guarantee her forty hours of work—when there was no business. This concept later became known as pump-priming, the idea of creating work. It was terrific for the person who landed a job out of it, but it could be miserable for the employer who shelled out money from profits that weren't there.

There was also Roosevelt's much-maligned WPA (the Works Progress Administration), which created so-called make-work jobs. The cartoonists of the day always showed the guys leaning on a shovel or falling asleep on the job. But at least we got some roads and bridges and parks out of it. And it sure beat drawing "relief" checks—an old-fashioned word for welfare. In those days, it was considered beneath a person's dignity to accept "relief" or charity. I don't want to be listed as a make-work advocate, but when I look at the deterioration in our roads and bridges today, I'm beginning to think we may just need another public works program.

By the time the 1936 election rolled around, I had become much more aware of presidential elections. That was the year Roosevelt ran against Alf Landon. I was twelve years old and in the seventh grade. Miss Allen was my homeroom teacher, and she was so Republican it was almost sickening. Every day, she wore a big sunflower to school with a picture of Landon in the middle of it, since Landon was from Kansas, the Sunflower State. She tried to tell us, *ad nauseam*, why a Republican should be back in the White House, but since most of my teachers back then were relatively affluent (my, how times have changed!) while we were the ragged little immigrant kids, the message didn't penetrate too deeply.

Roosevelt, of course, got reelected. And he had his work cut out for him. The country was trying to build the economy back up but it was still sputtering. For a couple of years the sky began to look a bit brighter, but pretty soon, in 1938, things took a turn for the worse again. Then in 1939 the war clouds started forming and we were forced to bolster our industrial might in order to wage the Second World War. I guess that war was the last "good" war we fought, because FDR convinced us we would make the world "safe" for democracy. How's that for brainwashing?

We went on fighting hard until 1945, and then we finally said we've got to stop this nonsense once and for all. And so we dropped the two really big bombs, and the war was over.

We had suffered through twenty years of hell and finally had our butter and our gas back (and our nylons and Hershey bars). Now that the smoke had cleared, we had to rebuild the world through our philanthropy and our generosity, which would take a lot of energy and money. Enter the Marshall Plan and the rebuilding of Japan.

We were fortunate to have extraordinarily active Presidents in the twenty years between 1932 and 1952. One stayed too long and died in office. It's often said that Roosevelt was too liberal. He sure did a lot of tinkering. Some of it failed and some of it worked, but thank God he was an activist. If he had been an ideologue, the Republic would have gone down the tubes. Presidents, of course, have to be right for their times, and, like FDR or not, he sure was.

And then we were blessed with Harry Truman, one of the best we ever saw. The little haberdasher from Independence, Mo. Old Harry was an idol of mine because he told it like it was. He didn't suffer fools easily. And he sure knew how to speak in clear, understandable English. He said, "The buck stops here." (And he said it first.) Like any good leader, he insisted, "Hold me accountable. Tough decisions are part of the territory." And, "If you can't stand the heat, get out of the kitchen." He was literally and lovingly "give 'em hell Harry." One line I'll always remember about him was when he said, "They call me give 'em hell Harry. But I don't give people hell. I just tell them the truth and they think it's hell!"

When some modern-day commentators compare Reagan with Truman, I want to retch. Truman became a victim of history when FDR died, just as Coolidge did when Harding died. But, wow, what a difference. Harry couldn't sit around silently and preach: "Let events run their course and everything will turn out okay." He had a world war to end, and he did it. He was a doer—and we should all be thankful he was.

By 1952, I was all of twenty-eight years old, had finished my education, and had been out in the real world for six or seven years. When I look back on those twenty years of my conscious life, they were the pits. They truly tested the mettle of America—and it flexed a lot but it never broke. There was so much turmoil that people got exhausted. They were completely burned out.

My theory is that in a democracy people finally get worn out from too much activism. They can endure only about twenty years of it at a stretch and then they need some time to reorient themselves and figure out whether they're going nuts or not. You can't have one war end in 1945 and a new one begin in 1950. That's a bit too much excitement for anyone to take.

So along came Ike, the conquering hero and our great father figure. The Marshall Plan was in place and all the mopping up was done. It was 1952, and we didn't need another man-made or natural disaster for a while; we wanted some peace. And so the country said, "Let's put this place on cruise control and take a sabbatical."

The years from 1952 to 1960, with Eisenhower occupying the White House, were our great rest period. The cycle was interrupted briefly by the Korean War and the Cold War in the final two years of that stretch, but generally speaking, it was a period of remarkable quiet.

We experienced hardly any inflation and essentially no unemployment. I think inflation was something like one half of one percent for the whole eight years, which I'd settle for any day. In world trade—get this—we were in debt to no one. There was hardly anybody even to trade with. It was pretty unbelievable how good we were feeling again.

But then we started to think we were entitled to all these good times. We were rocking and rolling all over the place. We had earned a well-deserved break and the hell with the austere times. Live it up, we told ourselves.

Even the cars we were building back then symbolized what was going on in our heads. We wanted to whoop it up a little, and so we built huge cars with those giant tail fins. We had gone through years of not buying any new cars at all. People clung to their old vehicles and held them together with Scotch tape. I know I did, and those cars were really unsafe at any speed. It was about time, we said, that we indulged ourselves again. We started the second-car family right then in the 1950s and 1960s. And why not? We were king of the hill. Even Elvis Presley and the rock-and-roll music of those years fit the mood perfectly (at least for the teenage crowd with strong eardrums). It was a time to put up your feet, relax, and enjoy. And, boy, did we ever.

Then the next cycle kicked in. After that eight-year pause, the dark clouds started forming once again. People felt that the country had been sleeping for eight years; it was high time to wake up. In came the magnetic Jack Kennedy. Right or wrong, he was going to embody the new youth movement. He said it was time we got America moving again. There was a new frontier out there.

Those of us who were sleeping too deeply received a few shots to help shake us out of our slumber. There was, of course, the big shot early on with Cuba. That jolt was followed by the success of the Russian space program and the intensifying feeling that we had become second-class citizens. Watching that happen ticked us off, and so we resolved to pour tons of money into space and one-up the Russians simply by putting a man on the moon. We returned to a real hands-on government in Washington. And then there was the worst shot of all—the one that killed our President.

And thus began the cycle of 1960 to 1980. For those of us born in this century, I have to think that the twenty-year period of 1932 to

1952 was far and away the roughest period of history we would ever have to endure. A Depression, World War II, Adolf Hitler, and an atom bomb left little time for the pursuit of happiness. If I had to go through a second period like that, I'd be looking for another planet to move to. Yet, to be honest, I didn't really know how miserable life was back then, because I was still so young and full of vim and vigor. When you're that age, problems tend to roll off your back.

Little did I know that the years 1960 through 1980 would challenge the record for misery. The Cuban crisis. The Kennedy assassination. The dastardly Vietnam War, the war we tried to kick under the rug as if it never even took place. A dispirited LBJ choosing not to run for reelection.

Then along came Nixon, who vowed he was going to put a stop to all our woes. And for a while, Nixon did what appeared to be a pretty able job. He put in wage and price controls and was extremely active. But then, of course, he failed us with Watergate. Nixon trudged off in disgrace, and the country really got rattled. It was awful enough to have a President shot in office by a kook in Texas, but when a President had to resign, we began to wonder about our very foundations.

Nixon happened to pick Jerry Ford as a caretaker. That wasn't all bad, because we might have been stuck with Spiro Agnew, except that he was picking up groceries for nothing down in Baltimore. And we managed to scrape through the final years of the cycle under Jimmy Carter, who wanted to teach us sacrifice and conservation and human rights at a time when we were getting a little tired of the malaise of the '60s and '70s. Hell, we wanted to feel good again.

That brought us to 1980, when the country welcomed a movie star into the White House. We were ready for an eight-year recess, and who better to lead us than a master of movie make-believe. If nothing else, you have to give Reagan credit for one thing. He told us exactly where he was coming from. Right off the bat he declared, "Life needn't be this way. We can go back to the good old days— maybe the fifties, or maybe the Roaring Twenties, because they were even better."

He promised he was going to make us a stronger nation and cut our taxes at the same time. He didn't know who was going to pay all the bills we ran up, but he said there was a curve floating around out there by the name of Laffer (some laugher!) and somehow it was going to take care of the bill collector.

The country listened to him talk and watched him smile and voted for him by landslides both times he ran. Not because we really believed those lines, but because we needed a long coffee break (but for eight years?). We needed someone like Reagan, just as we had needed someone like Silent Cal and someone like Ike. We had to have eight years to check our oil and get a lube job.

I suppose we got what we paid for. We can't get too mad that Reagan didn't know what was going on. We never felt he was very involved, anyway. In fact, we didn't want him to be too involved. He said he would keep his hands off and let us all do our own thing. He made us feel good again. Unfortunately, we never considered what the hangover was going to feel like.

It was the reelection of Ronald Reagan that prompted me to work out my Twenty and Eight theory. When I reflected on that tremendous vote of confidence the country had given him, I thought to myself: *But he hasn't discussed any gut issues yet.*

After four years in office, Carter had been pursuing what he felt was right. He was talking human rights and he was living through two oil embargoes and he was trying to get back hostages from the radicals in Iran. Who could feel good about those kind of things? We were ready to make our Twenty and Eight change, ready to hear Reagan say, "I want a strong defense so we don't have to worry about the Irans of the world anymore and I'll lower your taxes while I'm doing it." Nobody wanted to question his math.

When he and Walter Mondale squared off in those two televised debates, Reagan never said a word about the budget crisis. There wasn't a peep about the trade deficit. There was no talk about the Japanese. And he got away with that silence. I thought: *The public doesn't want to hear yet, dummy. All that stuff is the bad news. A guy sitting in his living room watching the tube has his own set of problems.* I realized then that we wanted to feel good, period. We wanted to believe everything was terrific, whether it was or not. And that's why the snake oil sold.

Alas, the last couple of years under Reagan have been pretty miserable. I had hoped he would muddle through the lame-duck years better than he has—but he's wobbling on two extremely lame legs right now. And it's on this sour note that we head into our next cycle.

Feast or famine is what these cycles are, and famine time is coming up. "Pain" is the name I've given to the next cycle. If my Twenty and Eight theory holds up, it's going to run all the way until 2008. I'm not thrilled to death that I'll have to hang on until I'm eighty-four just to see another calm period—but who am I to complain? After all, it serves me right. I came up with the theory, didn't I?

I think the years 1988 to 1992 will constitute the critical stretch, because they'll undoubtedly set the mood for the whole twenty-year cycle. From the way things look now, this next one could be a real beaut. We're sailing straight into uncharted waters. We're going into the cycle worse off than we went into any of the previous ones. Think about it. We've never before been a debtor nation. We've never before had such a polarized society. We've never before had so many homeless.

In fact, each down cycle seems to get deeper and more insoluble. Does that mean we're going to see the end of the Republic some day? Will we have another big war? I hope not, but I'm no longer so sure. Some days I feel that the balance sheet for the country looks so bad that if Moody's were to rate the U.S. government, it would be no buy. "Hey, don't loan any money to those guys," they'd say. "It's too risky."

If my theory's right, the bad times are going to strike fast and furious. That means that the next few years will witness some sort of a crisis or a crash. I don't know if all the banks are going to go out of business or if we're going to start a shooting war, but something pretty bad is likely to happen. Who knows, maybe the seeds are being planted right now between us and the Japanese. Maybe China will decide that they've had it with the Japanese and join up with us, or maybe they'll jump ship and hook up with Russia, or maybe Western Europe will start drifting away. All I know is that I see the kettle boiling again.

My friends keep telling me I shouldn't worry too much, but I'm getting a little cynical in my old age. My own people tell me we could never have a depression again, because we've got so many financial controls in place. And big shooting wars are out because nuclear weapons are such a deterrent. (That's like saying AIDS will never become a health problem, because we've come so far in medical science.) Well, Mike Deaver, Ivan Boesky, and Ollie North all managed to get around the controls, so I'm not so sure.

According to Felix Rohatyn, the banks are really bankrupt if you take a close look at their loans. In effect, our banking system is kaput. Our farmers are in deep trouble. Our schools are in trouble. Hell, I'm not just worrying over nothing.

But I think that before the fever breaks, there's going to be one more rise in the temperature or the patient isn't going to get well; he's going to die. And I think that's going to happen at the end of 1988. Then, from 1989 to 1992, we're going to have a real downer. I know that sounds awfully pessimistic, but the sooner we start tackling our problems, the sooner we'll recover from that downer.

And remember this: Just because we're going into a period of turmoil again doesn't mean that we have to get a President assassinated or have a guy impeached or fight a short war. We can learn from the past, and we can find some leaders who can get their arms around our troubles and make sure we won't witness another 1931 or 1932.

I can already sense a shift taking place. The American people are saying that railing against too much government and going against Washington are wrong. We're willing to pay for some services again. Not necessarily the big spending programs of LBJ or FDR, but we do want a little bit more than we've been getting. Notice that we're moving away from real conservative Republicanism. We're not shifting to outright New Deal radicalism, but we are becoming a bit more involved with our homeless and our poor and sick. Hands-on activism is coming back into style, or maybe it's just coming back in our guilty consciences.

*   *   *

If we're going to endure these times without crashing into a tree, government has to play a bigger role. We've got to have more planning (oh, that dirty word) and controls to right our course. The next President must find a way to ease the polarization, because we don't seem much like a "United" States anymore—just a bunch of fifty states each doing its own thing. Yes, it's boom and bust time. One day the Oil Patch is riding high; the next day it's flat on its back. The Rust Belt is rusting away, but now it's making a comeback (I think). The Farm Belt is hurting bad, and we're praying for good weather. Our high-tech areas, the Northeast and California, have been enjoying fat defense contracts, but they are in for some big cutbacks.

These problems were man-made—in Washington. Trade policy, farm policy, energy policy, and defense policy affect our lives for better or worse but more than we care to admit. Good policy is supposed to smooth out these boom or bust cycles. But our passive attitudes of the past eight years have exacerbated the problems. We're in for some gut-wrenching decisions in the next four years.

And so the odds are overpowering that we're going to have a very activist Democratic President. I don't think the Republicans can make it. People want a change, and I think they'll vote for the Democrats if they put up Mickey Mouse (or even Minnie). But if we should wind up with someone weak, as Gary Hart would have been, then we'll have to go to church right away and start praying he grows into the job.

Because strong leaders do grow into the job. When Roosevelt was first elected, a lot of skeptics huffed, "Who is this guy? He's an aristocrat—he shouldn't be a Democrat. And he's a cripple to boot." They said he'd make an awful President. But the times molded him. I don't think he had any strategy at first. All he knew and told us was: "The joint's on fire, bring in the water brigades."

Yet our current administration has refused to adapt to the times. That's the trouble with ideology over reality. Just when the people in Washington need a microscopic lens, they've been putting on the kind of glasses you wear while watching an eclipse. And unfortunately, the eclipse could be our own.

It's activist time again, and I think the next occupant of the Oval Office is going to get banged up pretty badly. After one term, he's going to have bandages from head to toe. It may be that we'll have a great one-term President. Or we may elect another Hoover, who will come in for a few years, just trying to make do, and get run over. If that happens, believe me, in 1992 we'll get another Roosevelt—or a near-dictator, benevolent or not, who'll really grab hold of the throttle and finally stop all the bullshit.

If what I've been saying is starting to depress you, you can easily decide I don't know what I'm talking about and go on your merry way. I'm certainly no historian, but this cyclical behavior has always existed. It wasn't just invented because I was born and somebody upstairs wanted me to antagonize people later on in my life by writing about it.

In fact, the great historian Arthur Schlesinger, Jr. (and his father before him) asserted that cycles go back to the early days of the country. In his recent book *The Cycles of American History*, he concludes that this pattern has persisted for a couple of centuries. To be sure, he views events from a somewhat different standpoint than I do. He defines the cycles in terms of shifts between a dedication to "public purpose" and a withdrawal to "private interest." His feeling is that there's a burst of governmental energy every thirty years, which alternates with the conservative restorations of the 1920s, 1950s, and 1980s.

Whatever makes events turn out this way, I doubt we can stop it—short of saying let's try a whole new order and start over again. Since that's a pretty radical approach, and in my generation, at least, we were always taught to work within the system, the only option we have is to try to master the cycles rather than overturn the whole system.

What I'm doing is sitting here and reading tarot cards. And the best advice those cards give me is for everybody to get active and involved.

Remember, in the annals of history there are only three things that galvanize a people: a financial panic, a real hot shooting war, or a plague. Do you know that on October 19, 1987, the market crashed and fell faster than at any time in our history, the same day we bombed an Iranian oil rig and heated up the Persian Gulf mess, and the newly appointed AIDS commission was having a hell of a time even defining the disease and was saying we may just have to wait and see who dies. If it's true that Americans are only galvanized into doing something when they are confronted with a real live crisis, then we should be primed for action; hell, we got all three in one day!

You can't afford to rest on your oars for the next four years. Everyone had better start rowing. And I mean more than just vote. Ask to serve on a commission, at either the local, state, or national level. Participate in the dialogues and debates. Go on local TV stations if you can and argue some of the issues that affect you. Dash off letters to the editor. Don't sit around and moan and slap your head. If your guy doesn't win in the next election, support to the hilt the one who does.

Don't ever underestimate the collective clout of an aroused people—especially if they happen to be Americans. How many times have you heard—or said—"Hey, I'm just a little guy. What power do I have? Who would listen to me?" Alone, maybe nobody. But together, you're a force so powerful you make or break democracies. So let's close ranks and get involved. Try to do something to move the needle. That's what I'm planning to do.

XVI

IF I WERE PRESIDENT

I grew up in an era when we were taught that any little boy in America could grow up to become President. I think that's still true—but only if you include little girls too.

Conventional wisdom says that if a peanut farmer and a movie actor can make it, anyone can. Even an auto executive or a used-car salesman. But I think it also tells you that we're getting a little hard up.

I've had more than ten thousand letters telling me I should go for it. I've heard from two Roosevelts and one Rockefeller asking me to run. I've had inquiries on my availability from labor unions, the NAACP, the United Jewish Appeal, and every Italo-American group on the face of the earth. Edgar Bronfman of the House of Seagram had me to lunch to offer his support. He really touched me, because he said his kids wanted me to run more than he did.

I've had two groups launch campaign committees on my behalf and have had to formally request the Federal Election Commission to make them stop. A former Republican governor put me on the New Hampshire ballot as a Democrat. The attorney general of Maryland put me on their ballot as a Republican. I have campaign buttons saying "Make a Date in '88 for Iacocca" and "I Like I". I have no less than six different bumper stickers promoting my candidacy, one of which is a bright-orange Day-Glo number trumpeting Iacocca/Nunn in '88. (That one originated in Georgia—where else?) I even have a picture of Mario Cuomo touting me by wearing a big lapel button with my mug on it.

As I've said for four years now, I'm flattered, and I already have lots of memorabilia to show my grandkids some day. But I'm going to have to explain to them why I chose not to heed the call to a higher duty. Believe me, I've heard all the reasons why I should: "You owe it to the country that gave you so much"; "We need a businessman running things for a change"; "It's a turnaround situation and you know how to do that"; and even "It's your destiny."

Just when I thought I'd heard all the reasons, I got a phone call from a psychic healer in Kentucky. He said that the previous night the spirit world had moved into the future and revealed that I should run for the presidency. If I did, he told me, I'd be pitted against George Bush and would win with ease.

I guess that when you're hot, you're hot. And so everybody says, "Why not give it a shot?" Well, first and foremost, I have no desire to switch professions at this late stage in my life. And believe it or not, politics is a profession. To be good at it, you've got to live and breathe it. I'm good at what I do, but it's taken me forty years to get the hang of it. I'm also a great believer in skills not being transferable. In a nutshell, you've really got to want something to do well at it. My fire in the belly isn't so hot anymore.

Second, my mother, my kids, and most of my friends have advised against it. They know me and my temperament. I worry a lot. And the next President will have a lot to worry about, believe me.

Still, I didn't take the matter lightly. I put down on a piece of paper all the pros and cons and then made the irrevocable decision to pass up this opportunity to serve my country.

Having made the decision early on, I then had to respond to thousands of letters, address the issue in my speeches, and try to hold off the curiosity of the press.

At first, the more I said I would not run, the more they said I must be running. I tried all kinds of responses. I started with General Sherman—you know: "If nominated I will not accept; if elected I will not serve." All that brought was lots of cartoons of me sitting in a Sherman tank making the proclamation. Then I said I would like to be President, but only if I were appointed. Real smartass! That didn't work. I tried "The only thing I'm running for is my life." That didn't work either. Then I really messed up. I began telling audiences I wouldn't run because I'm not a politician. The audiences leaped to their feet and applauded like crazy. That's exactly what they wanted—a nonpolitician. I dropped that line in a hurry. Next, I appeared on the same program as Dr. Ruth. I touted her as my running mate. I said we'd make a terrific ticket—I'd tell them *what* to do, and she'd tell them *how*!

Then I got serious. I said the whole job is too adversarial. The Congress is almost unmanageable (and that's as it should be, I guess; the checks and balances were put in there for good reason). The people of the press can be miserable (but I guess under the First Amendment they're supposed to be miserable). Still, those antagonists I could learn to handle. The real ball-buster is the bureaucracy. That's a different story. The Constitution didn't say much about that, but it's so big and so entrenched I don't know what you do about it. I have my own little version of a bureaucracy at Chrysler, and it tests my will and stamina every day.

However, before I bowed out, I decided to get one last opinion right from the horse's mouth. Listen to what these horses said.

Thomas Jefferson: "I am tired of an office where I can do no more good. . . . It brings nothing but increasing drudgery."

John Adams: "At the midterm of my single term I felt that the business of the office was so oppressive that I could barely support it two years longer. . . . Had I been chosen President again, I am certain I could not have lived another year."

James Garfield: "Four years of this intellectual dissipation may cripple me for the rest of my life."

Woodrow Wilson: "The amount of work a President is supposed

to do is preposterous. . . . A man who seeks the presidency of the United States is an audacious fool."

Warren Harding: "The White House is a prison. . . . I am in jail. . . . I knew this job would be too much for me. . . . God, what a job!"

Hearing all that moaning, I had to conclude that either the job description includes too much or we need Superman. (Of course, if you're a hands-off man like Reagan, you'd probably say, "Hey, what a piece of cake this is." With all his problems, Reagan actually looks younger now than when he first took office.)

But for a hands-on man like me, I decided things are just too ungovernable down there. At some time during my first year in office I know I would go stark raving mad. You see, common sense is no longer our government's strong suit.

Nevertheless, during those moments of fantasy when I imagined that maybe I could do the job, I did some pretty serious thinking. And if nothing else, I'd like the guys who are running for President to listen carefully.

First, I'd put my team together early, and name them. I know this is lousy politics, but it's good administration. No cronies, no rewards to the big fund-raisers or those with the biggest PACs. Just a list of the best and brightest people the nation has to offer. What could be more important?

As the system works now, you campaign for your party's nomination until you're ready to drop dead. Then you debate the other party's opponent for three months. Then you stay up all night to see if Dan Rather and the CBS computer have declared you a winner or a loser. Then there's time out for Thanksgiving dinner with the family, then a couple of days to do your Christmas shopping, and then the Inaugural Ball. We used to give the poor guy till March 4 to take office, but the Twentieth Amendment shoved it up to January 20. How, with so little time, can you get your people in place and up to speed and make your list of priorities for your first hundred days in office? Remember, the public usually tolerates your actions for just about that long—a hundred days.

In fact, the more I think about it, the more I feel we ought to demand that a candidate give us some examples of the caliber of people he'd be asking to serve with him. Let me give you some hypothetical thought-starters.

Before I got to brokering the vice-presidential nominee at the convention, I'd announce that Sam Nunn was going to be my running mate (and not just because I've got a lot of leftover bumper stickers). He's got the smarts in foreign affairs and defense matters and knows how to handle Congress.

Then I'd prepare the "man to beat" list for all the key jobs. To qualify, they'd have to be people with a proven track record. I'd pick guys like Don Rumsfeld for Secretary of State, Jack Welch of GE for Secretary of Defense, Felix Rohatyn or Paul Volcker for Treasury, Peter Ueberroth for Commerce, Doug Fraser for Labor—or, even better, Fraser as the chief trade negotiator. (He does know how to negotiate—take my word for it.) Most of these guys would have to take a hell of a pay cut, but I bet they'd do it happily.

For Secretary of Communications, a new job, I'd ask Tom Brokaw. Before he got rich and famous he used to cover the White House. Now, I'd use him to cover Sam Donaldson. Seriously, in this electronic age you need a concise, no-bullshit reporter to help you communicate clearly with the American people.

There are probably another twenty-five jobs that are crucial—OMB director, the EPA head, and so forth. I'd fill these jobs with top-flight people in the country who could run—and I mean run—these kinds of operations, not just with some cronies I grew up with.

Even with good people in place, I'd get extra help wherever I could find it. I'd enlist Nixon, Ford, Carter, and even Reagan to help out. I'd call them ambassadors at large, or whatever, but I'd pick their brains clean.

Next, I'd schedule a press conference the first day of every month to tell the American people what the hell is going on. I'd ask for a summit with Gorbachev every year, like clockwork. I'd do the same with the new leaders in China. Japan is so important I'd hit them twice a year. If OPEC would let me in, I'd like to see those guys at their quarterly meetings to find out what they're trying to do to me. I'd meet with the leaders of Congress to discuss national priorities and what we're doing about them every month too. I'd practically sleep with the head of the Federal Reserve System, because I've read he's the second most powerful man in the United States—and I'd like to make sure I keep him that way. He could maintain his independence, but only up to a point.

After I'd done my homework, I'd write out in longhand, in twenty-five words or less, what each of our nation's policies is. That would include foreign and defense policy, followed by fiscal, monetary, tax, trade, energy, and environmental policy. The Big Eight, I call them. If you don't believe that's a tough assignment, just try to write out this administration's policies—and that's after eight years!

Then comes the hard part: the national priority list. Not only is it tough to draw up but it changes from year to year. Yet it's essential. In my business life I've always kept in front of me a hot list of ten priorities, which I address personally. I did that at Ford and I do it at Chrysler. At the national level I might need more than ten. But let's start with the critical ones.

Being a businessman, I've always spoken out on issues directly affecting my company and its workers. The price of gas and interest rates are the lifeblood of my business. The difference between $1-a-gallon gas and $2-a-gallon gas and the difference between 10 percent and 15 percent interest rates on car loans can make or break me. I have 1980 and '81 to prove it. Exchange rates determine the competitiveness of my cars against the foreign competition in the world markets, so they, too, are vital. Those subjects and the twin deficits of budget and trade are the only topics I've ever sounded off on. I have some deeply held views on other subjects, but I keep them to myself. If I speak out on other than economic issues, everyone assumes I must be running for something, so I like to tend to my own knitting.

But, you see, I, too, am a father, a taxpayer, a consumer, and—I even think—a good citizen. Since I've made it clear I'm not running for anything, I'd like to share my views on the broader issues of the day. I classify them in two major categories: life-and-death issues and quality-of-life issues. Some are faraway problems; others are right on the street where you live.

The first overwhelming priority for any leader of the free world has got to be cooling off the nuclear race. This is really life and death—to the whole planet, that is.

It's funny, but in all my years in business I've never once contemplated anything nuclear—not even a nuclear-powered car. (Once, a friend of mine did have a nuclear pacemaker implanted in his chest.) Yet last year three incidents, one after the other, gave me pause.

En route to Italy, I noticed that there was a case of Carnation milk powder on my plane. When I asked what it was doing there, I was told that it was for a friend's daughter-in-law. She was pregnant, and ever since the Chernobyl accident she was terrified about drinking the local milk. Since the winds from the U.S.S.R. had been blowing in the direction of Italy, the local residents feared that the cows might have been infected by radiation.

Then, once we arrived in Italy, we went to the co-op market in Siena. We were picking out some leafy vegetables for dinner when one of the shoppers, surprised at our selections, cautioned us that people were not buying any greens.

"Why not?" I asked.

"Chernobyl," she said.

The third incident truly blew my mind. In Tuscany, this beautiful area where civilization began thousands of years ago, I ordered a delicacy of the region at a restaurant one night. It was called "Cingali," or dried sausages made out of wild boar meat. Hearing my order, some hunters in the restaurant warned me that no one was eating the sausages this year.

I said, "You must be kidding. They're my all-time favorite. What do you mean?"

"Chernobyl," they said.

I thought: *Christ, not again.* The hunters told me that the boars had swollen heads and were acting peculiar. They said it had to be the nuclear fallout. These guys and their ancestors had been hunting in these hills for a long, long time. They were as old as the olive trees and nothing fazed them. And yet these farm people were now alarmed because of a nuclear mishap half a continent away.

When I saw the terror in these people, it really struck me that the leaders of the world had better get together and deal with nuclear armaments. Because even if the world doesn't destroy itself, we're all going to go crazy worrying ourselves to death.

We've got to reduce the stockpile of nuclear weapons, we've got to ban testing, we've got to verify that the other guys aren't cheating, and we've got to make sure proliferation comes to an end so a guy like Qaddafi doesn't blow our whole game plan. I know this is a bloody complicated subject, but I think by now everyone understands it's a no-win deal for the whole world. I can't believe Gorbachev or anybody else wants to build missiles, store them in silos, and know even while he's doing it that they're all going to end up as excess inventory. Unless he's nuts. On second thought, let's meet with Gorby twice a year.

You see, I'm not an ideologue; I'm a pragmatist. You might ask what does that mean? Well, let me explain. Reagan, a major-league ideologue, thinks in terms of freedom fighters in Nicaragua and evil empires and Communists lurking behind trees. He rails against big government. He believes in absolutely pure and free markets. He wants to return to the good old days of the past.

But is it really unthinkable to sit down with the Russians and see, even with our radically different philosophies, if we couldn't cool it a little? Maybe we could start trading with them. Maybe even show them how we helped Germany and Japan grow up after the war. After all, those guys were on the other side too, and look how well they did. Is it really impossible to cut the defense budget by 20 percent and still preserve the safety of the free world? The Grace Commission said 30 percent of the budget was waste. And former Navy Secretary John Lehman told me that was probably on the low side.

Then why don't we cut it? I don't think it's the $600 toilet seats and coffeepots that are ruining us, disturbing as they are. Instead, it's spiraling personnel costs and overruns on the part of the big military-industrial complex. Too many layers of management, and too many generals and admirals.

Is it a crime to ask Japan and Germany, in our own self-interest, to pick up more of the defense tab? The new buzzword is *burden-*

sharing. I used that in speeches four years ago, because I was smitten with a concept called equality of sacrifice that we had tried at Chrysler. For the life of me, I can't understand why nobody wants to make the connection between the trade war and the Cold War.

Representative Pat Schroeder of Colorado, who was running for President, is considered a radical for proposing a simple plan that would tax the difference between what we spend on defense and what our allies are spending. To keep the sea lanes open and NATO in business, Germany spends 3 percent of GNP on defense; the Japanese, 1 percent—and the United States, 7 percent. So, she says, you want to send in a $50,000 Mercedes to the United States? Pay the 4 percent difference, or $2,000. You want to bring in a $25,000 Toyota? Pay $1,500, or the difference between 1 percent and 7 percent. Then allocate all these revenues to the common defense budget.

Take the Persian Gulf problem today. I don't know how we got there, but the specter of the U.S. Navy's making sure Kuwaiti tankers can get safely to Japan carrying oil at $15 a barrel is a little laughable. It probably costs the U.S. $5 a barrel to keep the Japanese industrial machine going so that the Japanese can export their industrial goods back to the United States. That's really stretching common sense.

No matter how much you spend, or misspend, for armaments, they're only as good as the strength of the economy backing them up. You can't be a number one military power if you're a second-class economic power, and we're fast on our way to becoming just that.

So the second priority is jobs. Without jobs, we don't grow. And without growth, we can't pay for the guns—or the butter. Without growth, we can't give ourselves or our kids a decent standard of living, which is an even higher priority than education. Because if we educate the kids and there's still no opportunity for jobs, we're going to frustrate them and they might take to the streets. Better they stay stupid, if that's the case.

So how do we create jobs? The only way I know is to create incentives to grow: incentives for invention such as R & D; incentives for investment such as tax credits; incentives to produce and not just consume; incentives to save, not just spend. We need a consumption tax in this country so badly it's becoming pathetic. Not just to raise some revenue, but to compete in world trade. Why are we taxing income and investment anyway? We made a start in the right direction in 1986 but we still didn't address revenues and world competition.

Let me give you just one example of why our present system stinks. I gave a speech in 1982 called "The Level Playing Field." Since that time, the phrase has come to mean a lot of things to a lot of people. The President referred to it in his State of the Union message in 1985. Washington now calls it LPF, which unfortunately means it's been adopted as a buzzword. Now it's in such common use that I expect it to show up any day now in Funk and Wagnalls dictionary.

Originally the term was meant to connote unfairness in trade or, more simply: "We are getting screwed." It applied to currency differences and tax differences. In my business, it amounted to about $2,500 a car and was always my prime example of why competing had become so unfair.

I'll use a German example to explain what I mean. (The Japanese example is better, but since I'm always picking on the Japanese, I'll pass them up this time.) In 1985, the German mark was 2.2 to the dollar. In the fall of 1987, it was 1.8 and heading down to 1.6 or 1.5, which means that the mark got stronger by about 30 percent (or the dollar got weaker). The net result is that a BMW that sold for $30,000 in the United States went up to about $39,000. But it also meant that a Chrysler minivan—an absolutely world-class product, by the way—dropped by about 30 percent when shipped to Germany.

So far, so good. But the taxes on the BMW coming into the United States are only 2.5 percent. The taxes on the minivan are 10 percent to the Common Market and 14 percent to Germany, making the car go up almost 25 percent—or ten times the tax on the BMW. Of course, the German customer doesn't want to pay that stiff tax, so I have to take it out of my profits. My profit margins drop by a couple of thousand dollars, all because I'm forced to compete against different tax codes.

Obviously, rebuilding our industrial base and the jobs that go with it depends on tax and currency policies that give us a fair chance to compete. If you have peace (that's a strong defense) and prosperity (a job for everyone), what else is there? Well, there's something called quality of life, which allows you not just pleasure but peace of mind. This is not a luxury; it can also involve matters of life and death.

Let's start with the AIDS issue. This one is a tester for all of us—and I mean all of mankind. In the old days, if you were a leper we isolated you, slapped you in a colony with other lepers. With AIDS we can't, because there are too many with the disease. And that's what it is—a communicable, deadly disease. We've done the obvious, preventive things so far. We've shut down the notorious bathhouses in San Francisco and New York. We've tested and cleaned up the blood banks. We've even passed out clean needles for the druggies. Condom sales are soaring, and that's good. These, too, came out of the closet. When I was a kid, we'd go into a drugstore and yell, "A pack of Camels, please." And then, when nobody was looking, we'd whisper, "And a pack of Trojans." Now we yell for the Trojans and whisper for the cigarettes.

Abstinence is certainly one solution, but that may be like shutting down breathing. Monogamy clearly puts the odds in your favor. But no one wants to consider testing for the poor millions who already have the disease. I've had a lot of people say to me: "If I had it, I wouldn't want to know about it. I would then feel better mentally, and I wouldn't know if I was passing it around. I also couldn't be discriminated against—and I certainly couldn't be accused of indiscriminately passing it on."

It's Catch-22 time. As Enrico Fermi said when he successfully launched the age of nuclear fission: "The equation is exponential. It goes to infinity." Well, it's the same with AIDS, and just as deadly.

So we've got to cure it. We can't legislate it away. Hemophiliacs, homosexuals, and drug addicts—the so-called high-risk groups—probably get little solace from the fact that we've now set up a presidential commission to look into the matter.

This whole problem hit home with me when I lost a good friend to the disease—Stewart McKinney, a warm, wonderful human being who also happened to be a U.S. congressman. Stewart was a representative from Connecticut, and a Republican to boot, and yet he fought like a dog against tough odds to get Chrysler the loan guarantees—which is how I got to know him. All his fight and courage didn't help him lick his sickness. If he had lived, though, I bet he'd know how to tackle the problem.

I don't. I tell people that. But I'd put the best minds in this country to work on it with the same vigor we showed in putting a man on the moon.

Let's move on to what may be the most contentious and controversial issue of all time: abortion. Personally, I'm against abortion, but I don't think I'd want to impose my views or morals on another. Because what's more personal than abortion, especially for a woman? (I've never heard of a male, even a male legislator, who had an abortion.) But again, it's the economics that are at issue. Why should a taxpayer pay $1,500 for someone else's personal decision? One night at a dinner with a group of editors from America's top women's magazines, I suggested that although abortion was a moral problem for a lot of people, essentially it was an economic problem, and that if I were President, I would limit one to a customer. Well, they went bananas. They said that many of their readers have had six or seven abortions. I suggested they enroll those readers in birth control classes—and I mean quick. And I don't mean beginner or intermediate classes. I'm talking the advanced course!

Is it really asking too much to ban the sale of handguns—or Uzi machine guns, for that matter? Why must we preserve a guy's right to buy a deadly weapon out of a catalogue? The National Rifle Association propaganda that "guns don't kill people; people kill people" is a lot of hogwash. Kids don't go to school in Detroit with rifles strapped to their legs; they go with cheap little Saturday night specials that they picked up for a couple of bucks. Now the NRA is even backing Teflon-coated bullets which can cleanly penetrate a bulletproof vest. Why should a powerful special interest group get away with such malarkey? (And by the way, guys, don't bother sending me all those letters. This is America, and I've got the right to my opinion, same as you do.)

It's taken a courageous woman like Sarah Brady, whose husband, Jim Brady, was gunned down by an assassin's bullet intended for President Reagan, to make a difference; she's been doing a bang-up job as a spokeswoman for Handgun Control Inc.

Like handguns, there's a killer just as deadly—and just about as quick—called drugs. Is it too radical to make second-offense hard drug pushers die for their deeds? The teenagers who curl up and die when they use their products pay with *their* lives. Shouldn't we even out this ultimate sacrifice called death just a little bit?

Those who are against the death penalty cringe at such solutions. But how else do you deter these hardened street criminals? If drug-dealing weren't so heinous a crime and if it weren't growing by leaps and bounds, maybe you could kick it under the rug. But this one won't go away. Our emotions get stirred and people will demand even the death penalty for a child killer or a cop killer. On that basis, these dealers should really get it, because they're involved in both.

Is it too much to give a drunk driver one warning, then yank his license, and then throw his ass in jail? I know we don't have enough jail space, but try telling that to someone who's just lost a child to a drunk driver. While our politicians were debating the issue, Candy Lightner started MADD, Mothers Against Drunk Driving, and dramatically changed our tolerant attitudes toward drunks behind the wheel.

All these subjects are a matter of life and death, whether you're thinking on a global basis or right here at home in your own neighborhood. Notice they're all defensive measures, starting with national defense itself. But, there's more to life than worrying about crime and killing and even nuclear war. There is (or should be) a certain quality of life, maybe even a little dignity. So why not give everyone who wants an education a chance to learn? And why not give peace of mind to people with catastrophic illnesses by providing some kind of guaranteed health insurance?

And lastly, let me propose my most controversial suggestion of all. Let's draft every able-bodied man and woman (wow!) between the ages of eighteen and twenty-five for one year of military or public service. Yikes, I can hear the screams already. But please, before you head for Canada, calm down a minute.

I imagine this proposal alone would lose me most of the popular vote. But I feel it's in everyone's best interest. The military would give you the normal choices, such as the army, navy, air force, or marines, while public service would include the Peace Corps, health clinics, hospices, welfare work, environmental projects, and so on. What a wonderful discipline for our young people, and what a terrific way to become involved with the quality of life early on.

All I'd have to offer is low pay and tough working conditions. The work could be physical, mental, or just plain soul-searching. Since it would include all eighteen- to twenty-five-year-olds, it would be certain to interrupt their formal education for a year or delay the start of their careers. Believe me, I know of no better way to make the United States great again than by asking our young people to serve their country.

As Americans, we don't believe much in drafting anybody for anything unless there's a national emergency. We like to volunteer our services for a good cause, but only if the spirit moves us. Well, it's the spirit I'm worried about. I think we're losing our national will—and spirit. And that's our biggest crisis of all. Hey, kids, you can't have it both ways. You can't have the material things and the quality of life you'd like without investing a little. You can fly a fighter plane, or man a radar station, or plant some trees, or minister to people dying of AIDS, or work in a lab on toxic wastes—your choice. After all, in a lifetime what's a lousy twelve months?

I suppose it's easy to mouth off about some of these things. But I'm going to try to do more than that. I'm one of the twelve members—six from the political arena and six from the private sector—who have been named to the new National Economic Commission. The Commission is a creation of Congress whose mission is to help the next President get this country moving again. Our charge is to come up with an economic game plan that tackles our gigantic budget deficit but does it in such a way that it expands economic growth and spreads the pain equally. Wow, what an assignment!

We began our work in March and we're expected to hand in our recommendations right after the election in November 1988. It's going to mean putting in a lot of long days, but what could be more important? I hope I manage to contribute something. Because, if we do our work right, we can have a hell of an impact. And if we blow it, we'll embarrass ourselves early.

The key thing to remember is that whether you're for or against any of the positions I've laid out doesn't matter. These are gut issues that must be addressed one day, for better or for worse. If not, we're going to victimize ourselves with fear. If we fail to get a grip on them, I'm convinced that we're not going to have much of a future. Politicians would rather not debate these issues, let alone take a stand on them. They know that a lot of vocal minorities would just drown them out. But this paralysis is affecting our morale.

Already, a truly alarming despondency has taken hold of many of our young people. I'm not talking about simple depression that fades after a day or two. A growing number of our teenagers are actually taking their lives because they believe the world's problems are insoluble.

This really gets down to the crunch point. After all, our greatest asset is kids. There are, of course, all levels of shock to a parent. I used to worry about how I'd react if one of my kids got pregnant before she got married. Then I thought: *Well, that's not the end of the world.* After that I worried, what if the kid came home and said she was a junkie? That could start the end of my world. Yet as difficult as any of those problems might be, nothing could be more mind-blowing than your kid's coming home and saying he was contemplating killing himself.

Like many parents, I never had to give more than theoretical consideration to any of these serious problems. My kids always seemed happy—even a scolding by me didn't get them that depressed. I watched over them carefully, and I was very fortunate.

Some families weren't as lucky. I realized that when a young kid who used to live in our neighborhood paid me a visit. Her name is Kelly Good. When Kathi was a teenager she used to go skiing with

Kelly's older sister, Kim, and all of us would go on picnics and ski trips together.

Kelly was about twelve then. Although I didn't particularly zero in on her, she always seemed a bit odd. When I saw her skiing, I noticed that she wanted to compete almost to the point of being a little off the wall. Other kids would ski and hope they didn't fall on their heads. She would do somersaults in the air, one of the first hot-dog skiers I ever met.

As I found out later, she was having a lot of problems and began taking to the bottle. By the time she was fifteen, she was drinking pretty heavily. Besides alcohol, Kelly's whole life revolved around TV. But the tube wasn't doing her a helluva lot of good.

Well, I didn't see or hear much about Kelly for a few years. Then, in May 1983, Kathi got a call from Kim. Kelly had tried to kill herself in San Francisco by drinking a quart of tequila and then swallowing a can of Drano as a chaser.

Thank goodness, Kelly survived. She's twenty-six now, a lovely, mature woman, and has chosen to devote her life to helping other kids who go through similar ordeals. In fact, she's working with a "suicidiology" organization. I never even knew there was such a word. "It's been a long trip to recovery," she told me. "I'm still constantly reminded by the scars that are left in my throat. But now my goal is to provide others with hope so that they can get through this mental cancer and regain some purpose."

Why are teenage suicide rates rising? A number of kids have told me that they grew up lonely. After all, a babysitter can't tuck you into bed and tell you she loves you the same way a parent can.

According to Kelly, an additional important factor is the uncertainty of the nuclear age. Young people are more idealistic. They think of the world as it should be, and they don't understand people who might get a kick out of bombing one another. For my generation, technology evolved gradually as time went by and so maybe we understood it better—or thought we did. Or maybe we're just jaded old farts.

Another thing Kelly mentioned is that technology moves so fast that trying to keep up gives you a feeling of inadequacy. There's too much going on and everything seems to carry such high stakes. Will I get a job? Will my SAT scores be good enough? The pressures close in on teenagers and they feel they can't cope.

When Kelly visited me, she rattled off some statistics that really blew my mind. Every day, as many as 1,000 children attempt suicide. Every year, 6,500 to 10,000 of them succeed. Since the 1950s, the teenage suicide rate in America has tripled.

It's gotten so bad, she told me, that only the Swedes suffer a higher rate. And at least they have an excuse: In winter it's almost totally dark there and it's bitter cold, to boot. And so the aquavit flows like water.

I don't think teenage suicide stems from some awful gene that's passed on by heredity. But the result is a shame on our civilization. What's left if the greatest asset of our country—the youth who are going to inherit it—decide that it's such a mess they'd rather not participate? They don't just drop out of high school; they drop out of life—and, man, that's the ultimate dropout.

If watching that happen doesn't force us to tackle our problems, I don't know what will. Because if we don't do something, we'll end up losing more kids to suicide than to wars.

So what does this all add up to? Well, somebody's got to grab these things by the neck and shake them, and that's what our leaders are supposed to do.

A President is nothing more than a caretaker from generation to generation. He gets four to eight years to watch over the country. When his term ends, he has to look back and assess what his legacy was. He has to ask if he educated the people a little more or if he made them richer or if he made them fatter or if he made them feel safer—or if he helped make their lives a little happier.

When he leaves office, it's usually his legacy he's worried about. But how about our legacy? If the people are poorer and their kids' test scores have dropped and they're using more drugs and they're worrying all the time, then he's really done nothing in the scheme of history. And I mean nothing. Not only does he have to live with that failure, but so do we.

XVII
INTO THE TWENTY-FIRST CENTURY

I doubt it's ever been easy to plan for your demise, and it sure isn't getting any easier. I come from a family where my father thought he was immortal and so he died without ever bothering to write a will. As a result, my will is so detailed it's the length of a short novel. Despite what many people think, I know I'm not immortal.

Since I'm not going to be bunking at the White House in this lifetime, and since it's nearly gold-watch time for me in the auto industry (I knew my days were numbered when people started telling me the Mustang qualified as an antique), I've been doing a little thinking about the final stage of my life.

Yes, I've accomplished a lot, but luckily there is always a little unfinished business lying around to keep a person out of trouble. In my remaining years at Chrysler, I want to make sure my successors are in place and ready to take over. That's really my only responsibility. But I'm also interested in more mundane matters—like attaining 15 percent of the car market and 25 percent of the truck market. And, of course, I want to keep our balance sheet healthy and see that everyone associated with the company prospers. I also have a lot of work left to develop our aerospace and financial services businesses.

If we at Chrysler do all that, I can see us giving GM and Ford a run for their money. And if the yen stays put, I can see us giving everybody else a run for their money. Beyond that, I don't want to do anything too gigantic. My latest vision of Global Motors is smaller than it was a few years ago. I no longer have this burning desire to put together a company that would be bigger than General Motors.

One thing I'm sure of: Chrysler won't suffer one iota without me. I'm totally convinced that life after Iacocca at the company will be just fine; it might even be better than life with Iacocca. I could kid myself if I wanted to massage my ego a little, but I know that there's really no customer loyalty to me. I've always said, "How loyal is anybody? If they go to a Chevy dealer and get a hundred dollars more on a trade-in, that's the end of their loyalty."

As I clean out my desk drawers in Highland Park, I may get a little misty-eyed but I doubt I'm going to miss the perquisites of power. Getting addicted to all the glory can leave you with a terrible downer if you're not careful. Being waited on and having your royal ass kissed for so long by so many people can leave you a little helpless. I'll never forget what happened to my old mentor Charlie Beacham. Shortly after he retired, he went on a pleasure trip. When he got to the hotel he was staying at, he paced around in the lobby for about a half hour waiting for someone to check him in. When no one showed up, he asked the room clerk what the procedure was. (Then, a few days later, he left, not knowing he was supposed to pay the bill.) For forty years, somebody had always done those things for him.

When you're high up the corporate ladder, you really do get accustomed to people taking care of you, even for the most trivial things. In the auto business, someone always gasses and washes your car every day. Once you retire, I guess pumping your own gas shouldn't be too complicated. But, you know, I'm not really too sure.

Over the years, I've watched a lot of guys grow old and turn into vegetables. You've got all these old turnips sitting around who were once running our largest companies. While they're still at the helm, they all say, "Oh, it'll never happen to me." But it does, because they forget to plan ahead.

I'm going on record now and saying it really will never happen to me. (And I hope I'm right.)

I do know that I'm tired of putting on a damn tux and going to functions, 90 percent of which feature me as the speaker. I've been keeping track: By now I've given 104 speeches at the Waldorf-Astoria Grand Ballroom alone. That's plenty for one lifetime. For forty years I've been wearing out my vocal cords in that ballroom, and I think it's time I gave the waiters a break. They must be awfully sick of my jokes.

I don't want to fly anymore, either. According to the airlines, there's a fatality every 2.1 million miles. Well, I long ago passed 2.1 million, so I've got to watch my step.

I'd also like to try to slow down another notch. Whenever I bump into Herb Siegel, the head of ChrisCraft, he kids me, "Boy, you look awful. You're working too hard. I told you twenty years ago, the meter's running, the meter's running. It just keeps clicking away. You better have some fun while you still can."

I guess I'm not built for too much leisure. But you have to face up to the reality that you don't live forever, and so you should try to enjoy life. A lot of times when I wake up and look in the mirror, I think: *What am I worrying about? I don't have that much to worry about anymore. Yet here I am, still at it.*

As I've gotten older, I have become a little more tolerant of situations and people. I try not to get frustrated now with things I know I can't change. There's a mellowing process in life—and I'm into my mellow years. Maybe as you get older you get tired. I guess they call it equanimity, when you're at peace with yourself. I'm pretty much there. I'm not restless, thinking there's something out there I haven't done yet. I don't want to conquer the world. I don't want to climb any more mountains. Hell, I'm even content to bypass the molehills.

On weekends my main objective is to try to relax. I'm not exactly painting the town red. I work my crossword puzzles religiously (the *Detroit Free Press* puzzle every morning and *The New York Times* at night), watch sports, and go for walks; I don't budget any time for homework these days, although I still read tons of mail. When my kids were growing up, I always used to put aside time for them. Now I put it aside for myself.

Once I finish up at Chrysler, I suppose I could load up the car and head for Florida, where I could sit by the pool and play golf everyday. But I'm not built that way. I have to work. What are you going to do with your life if you don't work? Not that I think I've slaved all my life. My hands don't have any calluses on them. But I do have scars on my brain—you just can't see them.

So my plan is to rub off some ointment on younger people. My intention (are you ready for this?) is to spend time teaching and preaching. I've been prattling long enough about the need for better education, so I'm going to see if I can practice it.

Kids are the future. That's why I spent the first part of my adult life raising them and why I'm going to spend the last part of my life teaching them. It's a lot more fun talking to the kids who are about to take over the world than talking to a bunch of old guys my age who've already had their chance and are on the way out.

I'll enjoy teaching, because I like young people. I think they keep you young, and I'd like to stay young mentally just a little longer.

Lehigh University, my alma mater, is where I'll probably hang my hat. Lehigh's the perfect atmosphere for me, since it's close to my roots. I still have family only five miles from the campus. By teaching there, I'll feel I'm paying back the college in a small way for all the good it did for me. But, more important than that, I'll be able to take all the education and all the experience of a lifetime and try to pass them on.

So I hope to end up on the mountaintop in Bethlehem as part of the Iacocca Institute. The mountain isn't likely to come to me, as it did to Mohammed, so I guess I'll just have to go to the mountain.

Campus life will be quite a different world for me. Unlike big-business executives, professors can truly enjoy life, although I hear that on some campuses, the bickering and the politics can get just as bad as in business. Still, at least campus life is supposed to be laid back. I may even have to go back to smoking my pipe.

I'm going to try to be a cross between a savvy, street-smart guy and an elder statesman. In almost every respect, young students today have it all over me. They're better educated. They're fresher. They have more energy. So what do I have to offer? Only one thing—I've been there. And I'm going to see if I can share with them a little about the real world.

Before I take a piece of chalk in hand or grab a pointer, let me first try to answer the question I get asked most by high school and college kids throughout the world: Who have been your own personal heroes, the ones you most admire and look up to?

I guess they always ask me these things because in 1986, according to the Gallup poll, Ronald Reagan was the most admired man in the world—and I had slipped into second place ahead of the Pope. These ratings have always struck me as a little silly. In 1987, when Gallup dropped me into an eighth-place tie with Gorbachev, it turned out that Gary Hart and Ollie North had jumped ahead of me. Now *that's* silly.

My heroes have a little more substance than these modern-day idols (including myself). When I was growing up, I had two heroes: Leonardo da Vinci and Joe DiMaggio. First of all, they were both Italian boys who had made it. Leonardo was a genius, the authentic Renaissance man. He was artistic (the "Mona Lisa"); he was a scientist (flying machines, jet propulsion, and parachutes); and he was a leading expert on anatomy. When as a kid I read his biography, I was also fascinated by his sense of humor. He said he really didn't like making love, but he did so only to study facial expressions that could later be used in his painting and sculpting. I thought: *Come on, Lenny, you're putting me on! Next thing you know, you'll be telling me you drink only for medicinal purposes.*

But seriously, it was the versatility of this man that first made me think a person can be as great as he wants to be. His mind and his hands literally let him reach for the stars.

Joe DiMaggio was a different kind of hero for me. He didn't reach for the stars, just the left-field seats, and gave me hope. He helped me shed a feeling I had when I was young that an immigrant kid—an Italian at that—would have trouble making it or ever earning respect. I can still remember pinning on my bedroom wall a newspaper sketch of Joe titled THE WALLOPING WOP. My reaction was a curious mixture of hurt and pride, but I promised myself I was going to grow up and be like Joe. I didn't know him in his glory days, but today he is still the quiet and gracious gentleman who excelled in his job and lived his life the same way. And what more can you say about an idol than "He didn't let me down."

Another of my lifelong heroes was Ben Franklin. He wasn't Italian, but at least he lived in Pennsylvania most of his life. Like Leonardo, he was a thinker, a tinkerer, a doer. When you think of Ben, you probably think of him flying a kite in a storm to prove his theory of electricity, or of him and the printing press. But in the *Encyclopaedia Britannica* he shows up in no less than eighty-one places. He appears under "Philosophy," "Literature," and "History of Science." He's under "Ballooning" and "Chess." (He helped make chess popular in the Colonies.) If you look up "Gulf Stream," you learn that he figured out what it was. (And it had nothing to do with the race track or our airplane company.) Under "Education," you find that he founded a college; under "Libraries," that he began the first library in the Colonies; under "Insurance," he established the first fire insurance company in America; under "Fire," he started the first fire company. And on and on.

By the way, he also developed the literary form called "autobiography." I guess I really have to thank him for that!

Of all the people in history, Ben Franklin is the man I'd most like to have a drink with; he'd drink port and I'd have a Dewar's. And of all the men who sat in that room and hammered out the Constitution, he'd probably be the least surprised by the amazing things that have happened since then. I don't think he'd fall off his chair when I told him we'd been to the moon. He was a philosopher and yet a practical scientist, always looking ahead. But above all, he understood what people can do when they pull together.

The Poor Richard Club of Philadelphia once presented me with its annual Benjamin Franklin award, which is one honor I really cherish. And, in a final twist of fate, the State of Pennsylvania through its Benjamin Franklin Partnership for Entrepreneurship just gave $3 million to Lehigh and the Iacocca Institute to help find ways to make America more competitive again.

So if a hundred years from now you see two guys flying a kite on the mountain in south Bethlehem, it'll be Ben and me still trying to figure out what makes things work.

I've already told you about Harry Truman. He was the greatest of our Presidents, and I've always tried to emulate him. He talked to the masses openly and in their own language. In a day when there was no TV, he had to do it the hard way, through newspapers and radio and the back of railroad cars. He was feisty, blunt, and very, very decisive. Most of all, he was a man of the people, who stood for basic American values.

Much as I admired him, I must confess that in 1948, I bet Mary that Tom Dewey would whip him. She gave me 15–1 odds and we bet ten bucks. "The country's going to elect him," Mary said, "because he's one of them." She was absolutely right.

My all-time favorite hero, though, is Winston Churchill. As you know, I'm big on communications, and to me the great communicator is not Ronald Reagan but old Sir Winston. He was eloquent and he was a leader, if only because he had such a command of the English language. During the Second World War, who else could have rallied the people to victory with exhortations such as "We shall fight on the beaches, we shall fight on the landing grounds, we shall fight in the fields and in the streets, we shall fight in the hills; we shall never surrender"? After the war, who else could have articulated the Russian threat more clearly and coined the expression *Iron Curtain*? His keen mind and quick wit separated him from all other leaders of his day. And he probably never heard of a cue card.

Churchill was also one of the funniest and bawdiest guys who ever lived. The story I like best—and there are hundreds of them—is the one about his going into the men's room in the House of Commons and spotting Clement Attlee, Britain's postwar Prime Minister, standing at the far end of the urinal. Churchill positioned himself at the near end.

Noticing the gap between them, Attlee said, "Being a bit standoffish today, aren't we, Winston?"

Without batting an eyelash, Churchill said, "Every time you see something big, Clement, you want to nationalize it."

But he was more than just quick. He gave life to the phrases *British pluck* and *true grit*, and he even made smoking long Havana cigars in public fashionable—God love him.

All of these men shared certain attributes. They were all pretty cool, showed grace under pressure, and didn't take themselves too seriously. In talking to students, I'd certainly suggest that they emulate these men and their qualities. I sure have. But then I'd tell them to get mad. Because only angry people change things—people who get mad enough to say, "Wait a minute, I'm just not going to stand for this."

While our young people have had their noses in their schoolbooks, we've been doing an absolutely miserable job of managing and moderating some of the violent economic changes taking place. My advice to them is to get mad about it. I don't mean riot in the streets. But get mad enough to demand the policies you need to compete in the world.

Get mad at the people in Washington who are burying you under a dungheap of debt. Tell them: "No more."

Get mad at the ideologues who want to make you martyrs to some eighteenth-century trade principles that everyone else ignores. Tell them: "I want a fair shot."

Get mad at anybody who tells you that you have to settle for packaged solutions. Tell them: "Get out of my way and let me think for myself."

That's how progress has always been made in this country of ours. People get mad and say "That's enough!"

Americans have always been a practical people. Until lately, we've always put common sense ahead of ideology.

When you think about it, we don't even have a fundamental ideology in this country. The genius of the Constitution is its tolerance of so many points of view and espousal of none. I don't think that's an accident. The Founding Fathers were too smart to try to tell us how to solve our problems; they just gave us a framework to work in.

And so the Constitution is not a blueprint, like Karl Marx's *Manifesto* or Chairman Mao's Little Red Book. It doesn't tell us what our goals should be, or how to reach them. It simply lays out the powers of the government and then lists a few basic values that we're not allowed to mess around with—like freedom of speech, freedom of religion, freedom of assembly. Beyond that, we're on our own.

Those rock-solid values—those basic freedoms—are really all you need. You can read the Constitution all day long and you won't find a single answer to the big problems we face today. There's nothing to tell you how to protect the environment, or what to do about terrorists, or how to put out a nuclear fire, or how to build the economy.

Which is just as well, because as Americans, we've often done a lousy job of figuring out just where we should be heading but we've usually done a brilliant job of getting there. Why? Because we aren't visionaries—we're practical. And because we've held on to those basic values.

We aren't leaving our kids a blueprint to solve their problems, but then nobody gave us one either. And if they had, it wouldn't have worked. You see, such plans generally work only for the people who invent or devise them. Basically, we were on our own. And guess what? So are our kids.

I talk to a lot of college presidents and professors, and I always ask them what college kids want most today. They usually answer: "Security"—a nice safe, secure, prosperous future—and there's nothing wrong with that. That's exactly what I wanted forty years ago.

But then I ask, "What's their biggest hang-up?" And they tell me it's the fear of screwing up—the fear of failure. Apparently, a lot of young people aren't too crazy about the idea of taking risks.

Well, you're never going to get what you want out of life without taking some risks. Everything worthwhile carries the risk of failure. I have to take risks every day. I'd rather not, but the world doesn't give you or me that option.

For us Americans, it all started with Christopher Columbus—he risked sailing right off the end of the earth. Maybe there is something in the Italian genes, but ten years ago when I went to Chrysler, I felt as if I were sailing right off the end of the earth too. And I'll tell you something, I think I came a lot closer than Columbus did! At least it sure felt that way when we were losing $6 million a day, every day for two full years.

But we came through it in good shape because we took some big risks. We didn't duck them; we faced them. And in the end we beat the odds.

And so I hope that you who are our future leaders don't duck them either. Other people have already taken most of the really big risks for you. Columbus showed you that the world is round—at least you've got a pretty good idea what's out there.

So just keep asking yourselves that big question, and keep asking it over and over again: "What kind of America do I really want?"

Up to now, you haven't had much of a choice. You've had to take the America that was handed to you. But now it's your turn.

Every generation of Americans has managed to leave the next one a little better off. That's part of our heritage. But each one leaves the next one a whole new set of challenges. And we sure are leaving you a couple of dandies!

But I'm not worried. Every time I get up before a bunch of our kids and get a whiff of all their energy and enthusiasm, I relax right away. Because I know that our country is going to be entrusted to awfully good hands. Forty years from now, when they're old fogies like me, ready to pass the torch on to another generation, I have no doubt at all that it will be burning brighter than ever. In fact, I'd bet my life on it!

INDEX

A

ABC (American Broadcasting Co.), 15, 116, 152, 200
Abortion, 294
Accountability
  in business management, 85–86
  children-parent, 16, 148
  in government, 209, 273–74
Adams, John, 285
Adams, Sherman, 173
Adolfo for Men fragrance, 60
Aerospace industry, 118, 121
Age. *See* Children; Elderly people; Young people
Agnew, Spiro, 161, 276
Agribusiness, 221
Agriculture
  auto industry, compared to, 225
  calamities, 228
  crisis in, 16, 219–29, 279
  declining farm population, 221
  diversification suggestions, 226
  federal subsidies, 16, 223–24, 226–27
  food surplus, 224–25, 227
  foreign trade and, 221–22
  Iacocca suggestions, 226–27
  importance of, 220
  Japanese trade barriers, 192–93
  productivity, 222
  real estate, 221, 225, 226
Agriculture Department, 221
AIDS, 67, 279, 282, 293–94
Airlines
  crash litigation, 132–33
  Iacocca on flying, 303
  quality of service, 250, 252
Alcohol consumption
  drunk driving, 295

  and teenage suicide, 298, 299
Allegis, 117–18
Allen, Miss (teacher), 272–73
Allentown, Pa., 5, 40, 56, 98, 157, 270
Allied-Signal, 102, 121
Allstate Insurance, 109
AMC. *See* American Motors
American Association of Retired Executives, 44
American Bar Association, 138
American Can. *See* Primerica
American Cancer Society, 109
American Chamber of Commerce, 193
American Motors
  Chrysler purchase of, 41, 83–84, 86, 101, 108, 122
Amerika (TV miniseries), 116
Angola, 227
Antitrust laws, 92, 122, 269
Arbitrage firms, 98
ARCO, 110, 200
Armed forces
  draft proposal, 295
  IQ level, 231
  See also Defense policy
Arms control, 290
AT&T, 243, 250–51
Atlantic Richfield. *See* ARCO
Attlee, Clement, 307–308
Atom bomb. *See* Nuclear weapons
Attorneys. *See* Legal profession; Litigation
Australia, 187
Autobiography, as literary form, 306
  See also Iacocca
Automation, 207–208, 251, 253, 262
Automotive industry, 220, 267
  car designs, 255, 275
  car loan rates, 288
  car prices, 258, 259, 260–61, 292

311

Automotive industry *(continued)*
  cost/quality issues, 252, 255–59, 260–64
  evaporative emissions, 263
  farm industry, compared to, 225
  foreign competition, 183, 188, 190–93, 195–96, 197, 245, 292
  fuel-economy, 213–14
  Japanese truck dumping, 175–76
  liability issue, 139–40
  plant safety, 261
  productivity, 222
  product quality, 251, 252–53, 257–59, 260–64
  recruitment, 245
  U.S.-Japanese arrangements, 118–19
  women in, 84–85
  See also Oil; names of specific companies
Automotive News, 65, 161

B

Back to the Future (film), 250
Baker, Howard, 175–76
Baker, James, 4, 144, 170, 194
Balance of trade. *See* Deficit, U.S.; Trade
Bangladesh, 222, 227
Banks and banking practices, 189, 269
  failures and problems, 98, 278, 279
  investment and arbitrage employment, 98
  Japanese ownership, 200
  See also Credit
Barnard, Christiaan, 155
Bartholdi, Frédéric-Auguste, 7
Beacham, Charlie, 78–79, 302
Belgium, 187
Bell System, 9
Bendix Corporation, 102, 104
Benjamin Franklin Partnership for Entrepreneurship, 306
Bergmoser, Paul, 6
Bermuda, 12, 32
Bethlehem Steel Corporation, 96, 246–47
Biden, Joseph, 144
Bidwell, Ben, 85, 191
Bierwirth, John, 105
Black Monday, 98, 111–12, 216, 217, 269, 281
Blanch, Ed, 25
Bloomfield Hills (Mich.), 21, 26
BMW, 292
Boesky, Ivan, 92, 94, 99, 279
Bok, Derek, 233
Bonds, Bill, 161
Bornemann, Winfried, 58
Bradlee, Benjamin, 146
Brady, James, 295
Brady, Sarah, 295
Brazil, 139, 177, 196, 213, 260
Bridgestone tires, 100, 101
Briganti, Steve, 6
Brock, William, 191
Broder, David, 146
Brokaw, Tom, 147, 149, 160, 161, 287

Bronfman, Edgar, 105–106, 283
Brown, Andrew, 26
Bryan, William Jennings, 220
Budget, U.S., 203–18
  balancing attempts, 207–208, 215
  defense, 177, 197, 199, 216, 244–45, 290–91
  entitlements, 210, 215, 216
  priorities, 210
  Reagan and, 277
  taxation and, 210–12
  See also Deficit, U.S.
Buick division (GM), 113
Burden-sharing (concept), 290–91
Burroughs-Sperry. *See* Unisys
Bush, George, 164, 166, 167, 284
Business. *See* Corporations;
  Industry; Stock market; names of specific companies and industries
Business Week, 111, 112, 161
"Buy American" appeals, 192, 258
Byrd, Robert, 125

C

California, 38, 171, 200, 217
Canada, 84
Capitalism. *See* Free enterprise
Capital punishment. *See* Death penalty
Caricature Society of America, 60
Carter, Jimmy, 169, 269, 276, 277, 287
Catholic Church. *See* Roman Catholic Church
Central Park (N.Y.C.), 13
CEOs
  delegation of authority, 78–80, 173–74
  employee relations, 77–78
  Iacocca's commandments for, 86–88
  letters to Iacocca from, 56
  press conferences, 151–52
  public relations, 115
  retirement and succession preparation, 85–86
  risk taking, 114–17
  salaries and perks, 110–14
  See also Management
Chad, 227
Challenger (space shuttle), 250
Chernobyl nuclear accident, 143, 147, 289–90
Cherry Sisters, 159
Chevrolet dealers, 120
Chicago (Ill.), 200
Children
  accountability, 16, 148
  affection, expressing, 32
  affluence and, 25–27
  American legacy, 309
  childrearing, 25–35, 117
  incentives for, 29–30
  letters to Iacocca, 55
  listening to, 28–29
  parental censorship, 117
  parental time with, 27–28
  rule setting, 30–31

INDEX

Children *(continued)*
  and Statue of Liberty fund, 9–10, 15
  TV viewing, 117, 236
  values, imparting to, 28, 30, 242–43
China
  agricultural production, 221–22, 227
  response to *Iacocca*, 55
  U.S. relations, 287
Chris Craft, 303
Chrysco, 118
Chrysler Aerospace and Technologies Company, 110
Chrysler Corporation
  American Motors purchase, 41, 83–84, 86, 101, 108, 122
  candor and litigation, 141
  car costs, 134–35, 260, 264
  computers at, 238–39
  corporate savings, 214
  corporate volunteerism, 110
  Dakota truck, 149–50
  equality of sacrifice concept, 291
  employee medical liability costs, 134–35, 260
  executive compensation, 110–14
  and export taxes, 211
  farm industry, compared to, 225
  fiscal crisis, 68, 69, 97, 158, 166, 169, 176, 203, 205, 212, 225, 294, 310
  fiscal crisis and corporate raiders, 97
  foreign competition, 292
  GM merger proposal, 121–22, 302
  and Goodyear Tire, 93
  graduate school recruitment, 245
  Iacocca priorities and goals at, 288, 301–302
  Iacocca retirement from, 302–303
  Iacocca succession plan, 85, 86, 301
  Japanese competition, 191
  Junior Achievement program, 110
  lawyers at, 127, 136–37
  Liberty project, 120
  litigation, 133–34
  Loan Guarantee Board, 169
  management-staff relations, 28, 77–78
  merger discussions, 100–105, 121–22, 302
  name-change suggestion, 118
  odometers on test-driving cars, 122–29
  pay scale, establishing, 241
  plant closures, 108–109
  plant safety, 261–63
  and press, 158
  quality emphasis, 249, 253, 256–58
  reorganization, 82
  retirement policy, 43
  robots, use of, 207–208, 251, 253, 262
  sales in Japan, 186
  staff and line, 75
  state loans, 176
  and Statue of Liberty project, 5, 11
  stock options, 111–12
  subcompacts (Omni and Horizon), 258
  tax impact, 211–12
  truck safety issues, 149–50
  TV commercials, 53, 62, 63, 64, 115–17
  warranty, 257
  women in management, 84
  workers' remedial education program, 233
Chrysler Fifth Avenue, 170
Chrysler New Yorker, 257
Chrysler-Plymouth dealers, 11
Chrysler Quality Institute, 256
Chung, Connie, 149–50
Churchill, Sir Winston, 65, 307–308
Cigars, 30, 66, 68, 308
"Cingali," 289
Civilian Conservation Corp., 272
Cold War, 197, 274, 291, 307
Colleges and universities. *See* Education; specific institutions
Columbus, Christopher, 309–10
Commerce Department, 287
Common Market, 292
Communications Department (proposed), 287
Communism, 290, 309
  See also Cold War
Competition and competitiveness
  business, and corporate raiders, 92
  foreign trade, 183–201, 292–93
  government policies and, 222
  with Japanese educational system, 234–36, 239, 242
  in product quality, 251, 252, 258, 259–61, 264
  social costs and, 264
Computers, 238–39, 251
Congress, U.S.
  administrative assistants, 147
  Economic Commission, 296–97
  federal deficit, 207–208, 216
  Iacocca on presidential relations with, 285, 288
  Iacocca and, 124, 156–57
Conscription. *See* Draft
Constitution, U.S., 136, 270, 285, 306, 308–309
  See also First Amendment; Twentieth Amendment
Consumer Reports, 252
Cooke, Janet, 148
Coolidge, Calvin, 268–69, 274, 277
Copyright laws, 196
Corn. *See* Grain
Corporate raiders, 91–106
  and business competition, 92
  Chrysler discussions, 100–105
  Chrysler-GM proposed merger, 121–22, 302
  examples, 93–95
  good vs. bad merger attempts, 101–105
  Iacocca on, 96–100, 121, 177
  rationale, 92–93, 97
  Reagan administration and, 177
  regulation suggestions, 105–106
  and stock market, 98, 106
  typical target, 97
Corporations
  budget-cutting, 209
  ethics, 98–99, 106, 110

INDEX

Corporations *(continued)*
  executive compensation, 110–14
  and graduate degrees, 235
  health care policies, 45
  liability risks and product development, 139
  mavericks in, 88, 114–15
  mergers and takeovers, 91–106
  name changing, 117–18
  public relations, 107–29, 141–42, 151–52
  quality and costs, 251
  remedial education programs, 233, 243–44
  research and development, 92, 101, 246
  retirement policies, 43–45, 261
  social responsibility, 108–10
  values, 98–101
  volunteerism, 109–110
  See also CEOs; Litigation; Management; Stock market; names of specific companies and industries
Cosby, Bill, 62, 245
Countess Mara ties, 259
Cox, Channing, 269
Credit
  car loan rates, 288
  consumer, 213, 216
  farm, 225
  foreign, 212–13, 214–15
Credit cards, 213, 216
Crime
  absence of (Liberty Weekend), 13–14
  drugs and, 295
  permissive society and white-collar, 99
Cronkite, Walter, 168, 245
Crosby, Phil, 256
Cruise, Tom, 62
Cuba, 275, 276
Cumberland (Md.), 95
Cuomo, Mario, 4, 10, 146, 284
Cuomo, Matilda, 12
Curran, Bill, 68
Currency rates
  and budget deficit, 218
  and farm exports, 221
  and trade imbalance, 194, 288, 292
  yen value, 144, 190, 194
Cycles, 267–82
  in American history, 281
  in presidential style, 268–78
  Twenty and Eight theory (Iacocca), 267–68, 277, 278–79
Cycles of American History, The (Schlesinger), 281

D

Dairy farming, 224
Daiwoo, 195
Dakota (truck), 149–50
Damone, Vic, 21
Dauch, Dick, 129
Death penalty, 295
Deaver, Michael, 170, 174, 279

Debt, federal. *See* Deficit, U.S.
Defense
  budget, 177, 197, 199, 216, 244–45, 290–91
  burden-sharing (Germany and Japan), 290–91
  draft proposal, 295–96
  policy, 233, 244–45, 277, 280
  U.S. allies, 290–91
  See also Nuclear weapons
Defense Department, 287
Defense industry, 114, 244–45, 280
Deficit, U.S.
  federal budget, 16, 203–18, 296
  personalization of, 204
  reduction options, 217–18
  trade, 183–89, 191, 204, 222, 277, 278
DeLuca, Ron, 64
Democratic Party, 270, 280
Deng Xiaoping, 55
Denomme, Tom, 75
Depression (economic), 215, 220, 269, 271, 276, 279
Des Moines Register, 144, 159
Detroit (Mich.)
  airport crash, 132
  press, 161
  school system salaries, 241
  streetpeople's plight, 227
  See also Automotive industry
Detroit Free Press, 161, 303
Detroit News, 161
Developing countries, 188, 215
Dewar's Scotch, 68, 306
Dewey, Thomas, 307
Diabetes, 32, 34
DiMaggio, Joe, 305–306
Dingell, John, 156
Dirksen, Everett, 110–11
Disadvantaged. *See* Homeless; Public service; Social conditions
Dodge Omni, 258
Dokkyo University (Japan), 16
Dole, Robert, 146
Dollar, value of, *See* Currency rates
Donaldson, Sam, 152, 287
Draft, military, 295
Drug abuse
  penalties, suggested, 295
  testing, 206
Drug industry, 139
Drunk driving, 295
Dukakis, Michael, 226
Dumping (trade term), 175–76
DuPont, 110

E

Eastman Kodak, 243
Eastwood, Clint, 245
Economic policies, U.S.
  Chrysler Corp. and, 115, 288
  free enterprise, 76–77, 140, 165, 177
  incentives for growth, 291–92

INDEX

Economic policies, U.S. *(continued)*
  National Economic Commission, 296
  Reaganomics, 207
  trickle down, 97, 206
  See also Budget; Trade
Edison, Thomas, 60
Education, 231–47, 279
  basic skills, 239–40
  career guidance, 236–37, 244–45
  career oriented, 244–46
  corporate programs, 233, 243–44
  dropouts, 232, 235
  equal opportunity, 295
  higher, importance of, 235
  Iacocca Institute, 246, 304
  Iacocca's daughters' experiences, 29
  Japanese vs. U.S. emphasis on, 234–36
  priorities, 233–34
  reading levels, 232
  specialization in, 238–39
  teachers, importance of, 241–44
  See also Reading; Teachers; names of specific institutions
Einstein, Albert, 60
Eisenhower, Dwight D., 173, 268, 274, 276
Elderly people, 42–46
  in Japan, 43
  letters to Iacocca from, 55
  See also Retirement; Social Security
Elections. *See* Government and politics; Presidential election of 1988
Electric power, 269
Ellis Island. *See* Statue of Liberty
Emissions, evaporative, 263
Encyclopaedia Britannica, 306
Engineers and engineering studies, 237–38, 244–45
Environmental protection, 260, 261, 263–64, 288
Environmental Protection Agency, 261, 263, 287
Escape from Sobibor (TV program), 116
Europe
  agricultural policy, 221, 222
  automotive trade, 292
  defense spending, 290–91
  See also country names
Ethics
  in business, 98–99, 106, 110
  in journalism, 148
Evans, Linda, 60
Everett's Pizza Parlor, 53
Exchange rates. *See* Currency rates
Executive compensation, 110–14
Exports. *See* Trade
Exxon, 200

F

Failure, youth's fear of, 309
Falkland Islands, 187
Family, as unit, 17, 23–24, 37–39

Farmers. *See* Agriculture
Farmer's Almanac, 221
Federal budget. *See* Budget, U.S.
Federal debt. *See* Deficit, U.S.
Federal Election Commission, 283
Federal Express, 251
Federal government. *See* Government and politics; Reagan administration
Federal Reserve System, 288
Fermi, Enrico, 293
Fields, W.C., 70
Fireworks (Liberty Weekend), 12, 13, 156
First Amendment, 158–59, 160, 285
Fleming, Ian, 103, 104
Fleming, Peggy, 11
Flemion, Robin (niece), 31
Fonda, Jane, 60
Food crisis. *See* Agriculture; Hunger, world
Forbes magazine, 161
Ford, Gerald R., 276, 287
Ford, Henry II, 82
Ford Motor Co.
  Chrysler Corp. competition, 302
  delegation of management, 78–79
  executives at, 27, 113
  foreign links, 196
  graduate degrees, view of, 235
  Iacocca at, 25, 82, 152, 288
  and Japanese dumping, 175
  merger attempts, 103, 104
  odometers on test-driving cars, 123
  Pinto suits, 141
  press relations, 152
  staff and line, 74–75
  public relations, 141, 150
Foreign aid, 198, 228
  See also Marshall Plan
Foreign investments, 212–13, 214–15
Foreign relations, 288
  with Japan, 16, 278, 287
  Reagan policies, 177
  U.S. adversaries, 196–97
  U.S. allies, 278–79, 290–91
  See also Cold War; Trade
Fortune magazine, 161, 173
Fox, Michael J., 250
France, 7, 12
Franklin, Benjamin, 213, 306–307
Fraser, Douglas, 112, 231, 284
Free enterprise system, 76–77, 96, 114 140, 165
Free trade. *See* Trade
F-16 fighter plane, 114
Fuel-economy standards, 213
Future Farmers of America, 220

G

Gallup Poll, 305
Galvin, Bob, 37
Garfield, James, 285
Gasoline. *See* Oil

Gastern, Carola, 57–58
Gatlin Brothers Band, 60
Geneen, Harold, 85
General Dynamics, 44, 114
General Electric, 257
General Mills, 109–10
General Motors
 and Chrysler, 84, 121–22, 302
 dealerships, 121
 management, 121
 media policies, 120–21, 150
 and Japanese dumping, 175
 Saturn project, 119–20, 177
 and South Korean cars, 195
Georgia, 284, 286
Gephardt, Richard, 156, 198
Germany
 auto exports, 292
 defense spending, 290, 291
 inflation (1930s), 271
Giants Stadium (Meadowlands, N.J.), 14
"Giggle Gang," 145
Global Motors, 302
GM. *See* General Motors
Golden Globe award, 117
Goldsmith, Sir James, 91, 93–95, 106
Good, Kelly, 297–99
Good, Kim, 298
Goodyear, Dick, 128, 132
Goodyear Tire and Rubber, 91, 93–95, 100–101
Gorbachev, Mikhail, 65, 124, 147–48, 287, 290, 305
Government and politics
 activism, 16, 279–81, 282, 308–10
 cyclical approaches to, 268–78, 278–81
 election turnouts, 212
 farm interests, 226
 farm subsidies, 16, 223–24, 226–27
 federal bureaucracy, 14–15, 16, 109, 163–64, 174–76, 285
 food policies, 227–28
 media and, 145–47
 and national pride, 251
 national priorities list, 288
 press, power of, on, 144–48
 public needs and, 209
 state, 176–77
 See also Congress; Presidential election of 1988; Presidency; Reagan administration; personal names
Governors, 174, 176, 177
 See also personal names
Governor's Council for Jobs and Economic Development (Mich.), 231–32
Governor's Island (N.Y.), 12
Grace Commission, 290
Graduate degrees. *See* Education; M.B.A.s
Graham, Katharine, 146
Grain, 221–22
Gramm-Rudman Act (1985), 207–208
Grant, Ulysses S., 269–70
Gray, Harry, 85

Great Britain, 307–308
Great Depression, 269, 271, 276
Green Bay Packers, 89
Greenbrier resort (White Sulphur Springs, W. Va.), 37, 156
Greenland, 187
Greenmail. *See* Corporate raiders
Greenwald, Jerry, 85, 123
Gridiron Club, 145–46
Griffiths, Martha, 232
Grumman, 105
Guidance counselors, 236–37
Gulfstream, 60
Gun control, 294–95
Guthrie, Janet, 84

H

Hahn, Jessica, 160
Handgun Control Inc., 295
Handguns, 294–95
Harding, Warren, 268, 274, 286
Harris poll, 241
Hart, Gary, 143, 144, 280, 305
Harvard Business School, 74, 78, 102, 235, 245
Harvard University, 233
Haverly, Michael, 9–10
Hawaii, 200, 217
Health care costs and insurance. *See* Medical costs
Hell's Angels, 9
Hennessy, Ed, 121–22
Hentz, Ned, 22, 23, 35, 66
Hillsdale College, 34
Hilty, Don, 224
Hiraizumi, Wataru, 199–201
Hitler, Adolf, 271, 276
Hodel, Donald, 4, 166
Hog Callers of America, 53
Holland. *See* Netherlands
Homeless, 16, 227, 279
Honda, 194
Honeywell, 110
Hoover, Herbert, 269–70, 281
Horizon, 258
Hostile takeovers. *See* Corporate raiders
Hot Tomatoes Dance Orchestra, 61
House of Seagram. *See* Seagram's
Hughes Aircraft, 121, 244
Human rights, 276, 277
Hummel figurines, 25
Hundred Days (Roosevelt administration), 271–72
Hunger, world, 222, 227
Hyundai, 195

I

Iacocca, Antoinette (mother), 4, 37–42, 45, 51, 67, 210, 284

INDEX

Iacocca, Delma (sister), 56
Iacocca, Kathi (daughter), 6, 21–35, 39, 48, 50, 52, 66, 68, 284, 297–98, 303
Iacocca, Lee
  on abortion issue, 294
  on accountability, 16, 85–86, 148, 209, 273–74
  on AIDS issue, 293–94
  autobiography, 9, 16, 54, 58, 62, 144, 157, 196
  on burden-sharing, 290–91
  on candor, 122–29, 141, 150–52
  challenge to youth, 308–10
  childhood recollections, 5–6, 98, 131, 221, 231, 270–71, 272– 73
  on childrearing, 24–35, 117
  on cigar smoking, 30, 66, 68, 308
  on communication, 73, 74, 80–81, 87, 158, 287, 307
  on computers, 238–39
  on corporate responsibility, 108–10
  on corporate risk taking, 114–17
  on corporate takeovers, 96–100, 121
  and crossword puzzles, 113, 241, 303
  cycle theory of U.S. history, 267–82
  on decision making, 75–76
  on deficit spending, 214–18
  on deindustrialization of U.S., 109, 244–46, 165
  on delegation of management, 78–80, 173–74
  draft/public service proposal, 295–96
  on drugs and alcohol, 295
  on education issues, 231–36, 238–39, 244
  on elderly people, 42–46
  on executive compensation, 110–14
  on fame and privacy, 66–67, 70
  on family, importance of, 17, 23–24, 37–39, 46
  on farmers' plight, 219–21, 223–29
  on federal deficit, recommendations to counter, 215–18
  on federal government, 14–15, 16, 109, 163–64. *See also* subhead Reagan and Reagan administration
  on Ford management style, 74–75
  on gasoline consumption, 214–15
  gifts sent to, 59
  on GM, 119–22
  on grief, 33–34
  on grudge holding, 65
  on handgun control, 294–95
  on his daughters, 6, 21–35, 39, 48, 50, 52, 66, 68, 284, 297– 98, 303
  on his earnings and perks, 111–12
  on his education, 236–37, 240, 243
  on his father, 5–6, 13, 17, 28, 29, 38, 39–40, 41, 67, 69, 98, 131, 221, 223–24, 231, 271, 272, 301
  on his future plans, 301–305
  on his health, 154–55
  on his heroes, 305–308
  on his mother, 4, 37–42, 45, 51, 67, 210, 284
  on his religious beliefs, 67–70
  on his second marriage and divorce, xii, 46–52, 59, 61
  on his wife Mary, 22–34, 46, 49, 61, 69, 165–66, 307
  honors and awards, 59–60, 306
  on Japanese auto trade competition, 190–201
  on lawyers and litigation, 134–41
  letters to, 9, 15–16, 55–62, 219
  on liberty, concept of, 15–16
  on loneliness, 46, 298
  management commandments, 86–89
  management style, 77, 78–79
  on mavericks, 88, 114–15
  Miami Vice appearances, xii, 54, 63–64
  National Economic Commission, 296–97
  on news sources, diversity of, 160–61
  on nuclear issues, 289–91
  personal prayer, 68–69
  philanthropy, 112
  on political activism, 279
  prescription for U.S., 16–17, 308–10
  on presidency (U.S.), own touted candidacy, 41, 146, 154–55, 283–85
  on presidency (U.S.), predictions on next, 146, 280–81
  on presidency (U.S.), recommendations for next, 286–96, 299
  on Presidents (U.S.), cycles in styles, 268–78
  and press, 146–62
  public response to, 53–66, 219, 305
  on quality, concept of, 249–58
  on quality, costs and, 259–64
  on Reagan and Reagan administration, 164–79, 277–78, 290, 307
  and Regan, 169–72
  on reading, 239–41
  retirement plans, 85–86, 302–305
  on school prayer, 233
  on South Korean work ethic, 196
  speaking engagements, 107, 138, 156, 160, 220, 245, 290–91, 292, 303
  speaking style, 65
  and Statue of Liberty commission, 3–16, 59, 62, 153, 155–56, 170
  Statue of Liberty–Ellis Island Centennial Commission, dismissal from, 3–4, 170
  on suicide, teenagers', 297–99
  syndicated column, 158
  on taped interviews, 153
  on taxes, 210–12
  on trade policy, 190
  trade policy proposal, 198–99
  on trade war–Cold War connection, 291
  on U.S. priorities, 289–96
  on values, 16–17, 28, 30, 98–101, 242–43, 245–46, 254
  on women in automobile industry, 84–85
  work habits, 112
  on young people's attitudes and problems, 98–100, 295, 297–99, 304, 308–10
  See also Chrysler Corporation

Iacocca, Lia (daughter), 6, 21–35, 39, 48, 50, 52, 68, 70, 284, 303
Iacocca, Mary (wife), 22–34, 46, 49, 61, 69, 165–66, 307
Iacocca, Nicola (father), 5–6, 13, 17, 28, 29, 38, 39–40, 41, 67, 69, 98, 131, 221, 223–24, 231, 271, 272, 301
Iacocca (autobiography), 9, 16, 54, 58, 62, 144, 157, 196
Iacocca Foundation (diabetes research), 34
Iacocca Institute (Pa.), 246, 304, 306
IBM, 45, 251
Icahn, Carl, 91, 96
Illiteracy. *See* Reading
Immigrants, 6, 8, 10
Imports. *See* Trade
Income tax. *See* Taxes
India, 222
Indianapolis 500, 84
Indonesia, 222
Industry (general)
  deemphasis in U.S., 109, 164–65, 244–46
  Lehigh Iacocca Institute on, 246–47
  quality concept, 249, 251–53
  See also Corporations; Japan; specific companies and industries
Inflation, 214, 251, 271, 274–75
Information explosion, 239
Insurance
  for the future, 253
  health, 44, 134–35, 260–61, 295
  Liberty Weekend fireworks, 156
  malpractice, 135
Interest rates, 214, 225, 288
Internal Department, 5
  See also Hodel, Donald
Interior Revenue Service, 206
International Food for Peace program, 227–28
IQ levels, 231
Iran
  contra issue, 78, 171–72, 172–73
  hostage issue (1979), 277
  Persian Gulf, 282
Iron Curtain (term), 307
Israel, 60
Italy and Italian-Americans, xii, 4, 60–61, 70, 167, 184, 283, 289, 305, 310
ITT, 85
Iwo Jima Memorial, 186

J

Jackson, Jesse, 145
Jackson, Reggie, 114
Jackson, Michael, 111
Japan
  auto exports, 188, 190–93, 245
  burden-sharing, 290–91
  Chinese grain sales to, 221
  Chrysler sales in, 186
  consumer goods, 194–95, 244–45, 260
  defense spending, 198, 199, 290, 291
  economic prosperity, 198–99, 260
  education, approach to, 234–36, 239
  elderly in, 43
  GM and, 119
  Iacocca speech in, 193
  legal profession in, 138
  management style, 76–77
  mergers, attitude toward, 96
  Persian Gulf oil, 199, 291
  product competitiveness, 259–60
  quality of products, 253, 255–56, 257
  real estate investments in U.S., 199, 200, 217
  rebuilding of, 273, 290
  research and development, 244–45
  and South Korean competition, 195–96
  teachers' salaries, 242
  technological development, 237–38, 244–45
  tire companies, 100–101
  trade practices, 156, 164, 175, 183–201, 259–60, 277
  truck dumping, 175–76
  and U.S. federal deficit, 212–13
  U.S. relations, 16, 199, 278, 287, 291
  workforce, 195, 198, 199, 255
  yen value, 144, 190, 194
Jeep, 64, 84
Jefferson, Thomas, 143, 161, 215, 285
Jennings, Peter, 147, 160
Job issues. *See* Unemployment; Workforce
John Paul II, Pope, 147, 178, 305
Johnson, Don, 63
Johnson, Lyndon B., 276, 279
Johnson, Peggy, 47–52, 59
Jones, Tom, 110–11
Journalists. *See* Press; personal names
Junior Achievement, 110
Junk bonds, 92, 106
Justice Department, 123, 127–28, 200

K

Kanehann, Bill, 240–41
Kansai International Airport, 185
Kansas City school district, 242
Kansas City Times, 149
Kennedy, Edward M. (Ted), 10, 146
Kennedy, John F., 60, 275–76
Kenosha (Wis.) auto plant, 108–109
Kissinger, Henry, 153
Koch, Edward (Ed), 12, 13
Kondobo USA Inc., 200
Korea. *See* South Korea
Korean War, 225, 274
Kristofferson, Kris, 116
Kristol, Irving, 164

L

Labor. *See* Workforce; Unemployment
Labor Department, 287

INDEX 319

Labor unions, 112, 165, 283
Landers, Ann, 143
Landon, Alfred, 272
Las Vegas (Nev.), 200
Lawyers. *See* Legal profession; Litigation
Layoffs. *See* Workforce; Unemployment
Legal profession, 135–39
   See also Litigation
Lehigh University, 84, 237, 244, 245, 246, 304, 306
Lehman, John, 290
Leiby, Jimmy, 243
Lent, 68
Leonardo da Vinci, 305
"Level Playing Field, The" (Iacocca speech), 292
Levine, Dennis, 99
Lewis, Joe E., 220
Liability. *See* Litigation
Liberty, Iacocca's personal concepts of, 15–16
Liberty project (Chrysler), 120
Liberty Weekend celebration, 11–15, 39, 155–56, 166–67
Libya, 14, 187
Liechtenstein, 223
Life magazine, 153
Lightner, Candy, 295
Likins, Peter, 246
Lincoln, Abraham, 68
Litigation, 131–41
   airline crashes, 132–33
   candor and, 141
   Japanese attitude toward, 138
   lawyers, 135–37, 137–38
   liability fears, 134–35, 139–41, 264
   libel suits, 159
   medical malpractice protection, 134–35
   odometer-test cars suit, 122–29
   Pinto (Ford) suits, 141
   prevalence, 131–34
   product costs, impact on, 134–35, 137–38, 264
   punitive damages concept, 138–39
Little, Arthur D., Co., 244
Little Red Book (Mao), 309
Lombardi, Vince, 89
Los Angeles Times, 5, 161
LTV, 96
Lutz, Bob, 85

M

MacArthur, Douglas, 76
MacLaine, Shirley, 69
MADD (Mothers Against Drunk Driving), 295
"Made in Japan," 255
"Made in the USA," 249
Magowan, Peter, 95
Malaysia, 183
Malpractice insurance, 134–35
Management, 73–89
   accountability, 85–86
   as art, 79
   brand management, 81
   chain of command, 80–81
   consensus, 76–77
   corporate raiders and, 92, 93, 97
   delegation of authority, 78–80, 88, 173–74
   employee relations, 77–78
   Iacocca's commandments of, 86–89
   Japanese style of, 76–77
   job switching, 170
   limit setting, 88
   mavericks, 88, 114–15
   planning policies, 89
   priorities, 86–87
   Reagan's style of, 173–74
   retirement and succession, 85–86
   skip meetings, 80–81
   staff and line, 74–76, 87–88
   women in, 84–85
   written communication, 87
Manifesto (Marx), 309
Mansfield, Mike, 193
Manufacturing. *See* Industry; specific companies and industries
Mao Zedung (Tse-tung), 309
Marcos, Ferdinand, 55, 147
Marcos, Imelda, 147
Marshall Plan, 198, 273, 274
Martin, Phillip, 65
Martin Marietta, 102
Marx, Karl, 309
Maryland, 283
Matsui, Bob, 156
M.B.A.s, corporate view of, 235
McDonald's, 232
McGovern, George, 227–28
McKinney, Stewart, 293–94
McNamara, Margie, 241
Meadowlands (N.J.), 14
Medal of Liberty, 10
Media. *See* Press; Television
Medical care
   costs and malpractice insurance, 134–35
   elderly, 44–45
   employee benefits and product costs, 260–61
   guaranteed, 295
Meese, Edwin, 170
Mercedes, 146, 149
Mercer, Robert, 93–94, 95
Mergers. *See* Corporate raiders; company names
Merrill Lynch, 93, 169
Mexico, 177, 196, 213
"Miami Nice" (municipal courtesy program), 257
Miami Vice (TV program), xii, 54, 63–64
Michigan, Governor's Council for Jobs and Economic Development, 231–32
Microchips, 188–89
Middlebury College, 22
Military-industrial complex. *See* defense industry
Military service. *See* Armed forces; Draft
Miller, Mary, 8–9

Miller, Steve, 85
Minnelli, Liza, 11
Miss Liberty. *See* Statue of Liberty
MIT (Massachusetts Institute of Technology), 84, 244, 245, 246
Mitterand, François, 12
Mobil Oil Company, 115
Model T Ford, 131
Mondale, Walter, 208, 277
Monetary system. *See* Currency rates
Motorola, 37
Mozambique, 227
Mozart, Wolfgang Amadeus, 60
Muldowney, Shirley, 84
Mustang, 301

N

NAACP, 283
Nagy, Jim, 23
Nakasone, Yasuhiro, 156, 164, 192
NASA, 250
National Automobile Dealers Association, 77
National debt. *See* Deficit, U.S.
National Economic Commission, 296–97
National Recovery Act, 272
National Rifle Association, 294
Native Dancer (race horse), 60
NATO, 291
NBC News, 144, 149–50
Nelson, Willie, 11, 14
Netherlands, 187
Neuman, Alfred E., 154
Newark (N.J.) auto plant, 261–62
New Deal, 271, 279
New Hampshire, 283
New Jersey
 Liberty Weekend, 11, 14
New Look magazine, 153
Newman, Paul, 62
Newsweek magazine, 158, 161
New York City
 Japanese real estate purchases, 200
 Liberty Weekend celebration, 11–15
New York *Daily News*, 124
New Yorker, The, 160
New York Philharmonic, 13
New York Times, The, 54, 113, 144, 161, 189, 200, 255, 303
New York World, 7
Nicaragua, 290
1986 (TV program), 149
Nixon, Richard M., 129, 143, 150, 178, 276, 287
North, Oliver, 172, 174, 279, 305
North Atlantic Treaty Organization, 291
Northrop, 110–11
NRA, 272
Nuclear power
 Chernobyl accident, 143, 147, 289–90
 dangers of, 289–90
 liability risk, 140

Nuclear weapons, 146, 276, 290, 298
Nunn, Sam, 124, 284, 287

O

Oakland University, 34
Occupational Safety and Health Administration, 261–63
Office of Management and Budget, 287
Ohio, anticorporate takeover legislation, 94
Oil
 embargoes, 213, 214, 228, 277
 fuel-economy standards, 213
 gasoline consumption, 214–15, 228
 gasoline evaporative emissions, 263–64
 gasoline prices, 288
 imports, 197, 199
 Persian Gulf, 199, 282, 291
 sales, 217
 taxes on, 216, 217
Old age. *See* Elderly people
Oldsmobile, 170, 172
Omni, 258
Omnicom, 118
O'Neill, Thomas (Tip), 169
OPEC, 197, 287–88
O'Reilly, Frank, 123, 129
Orpheum Wiener House, 272
OSHA, 261–63
Ozone, 67, 263

P

Pace, Frank, 44
Pakistan, 187
Palmah Tzova (kibbutz), 60
Paper industry, 260
Paraguay, 187
Paulsen, Pat, 154
Paulson, Allen, 60
Peace Corps, 296
Pennsylvania, 306
Pennzoil, 133
Pensions. *See* Retirement
Penthouse magazine, 153
People magazine, 60
Pepper, Claude, 43
Perot, Ross, 121, 122
Persian Gulf, 199, 282, 291
Petersen, Don, 103
Petroleum. *See* Oil
Philippines, 147
Pilliod, Chuck, 93
Pinto, 141
Pioneers of America, 9
Plymouth Horizon, 258
Pocatello (Ida.), 53
Poindexter, John, 170, 171
Polaroid, 243
Politics. *See* Government and politics;
 Presidential election of 1988

INDEX

Pollution. *See* Environmental protection
Poor Richard Club, 306
Poor Richard's Almanac (Franklin), 213
Portugal, 184
Potamkin, Victor, 121
Pownall, Thomas, 103–104
Presidential election of 1988
   aspirants and press, 146–47
   Iacocca and, 41, 146, 154–155, 283–85
   Iacocca predictions on, 146, 280–81
Presidency, U.S.
   cycles in past, 267–78
   Iacocca's advice for next, 286–88
   Iacocca on priorities, 289–99
   Iacocca's proposed Cabinet, 287
   leadership and legacy issues, 299
   past Presidents on, 285–86
   press conferences, 287
   problems facing next, 280–81
   See also Presidential election of 1988; specific names
Presley, Elvis, 14, 275
Press, 143–62
   business relations with, 151–52
   candor with, 151–52
   and Chrysler fiscal crisis, 158
   freedom of, 158–60
   Iacocca's expectations of, 154–56
   Iacocca's reading habits, 161
   and politicians, 146–47
   power of, 144–48
   presidential conferences, 152, 287
   public distrust of, 148–52
   and Statue of Liberty celebration, 7, 155–56
   taped interviews, 153
   See also names of specific publications
Primerica, 118
Princeton University, 135
Product quality, 249–64
   competition and, 251, 252, 258, 259–61, 264
   costs and, 259–64
   customers and, 252–53
   diminished, 249–51
   Japanese, 253, 255–56, 257
   liability cases, 134, 139–40
   liability costs, 134–35, 137–38, 264
   profits and, 251
   U.S. vs. foreign-made goods, 256, 258
   warranties, 252, 257
   workers' commitment to, 250, 251, 253, 254–56
Protectionism. *See* Trade
Proton, 183
Proxmire, William, 106
Public opinion
   on budget deficit, 208
   on Japanese workmanship, 255–56
   most-admired-men polls, 305
   on Reagan, 60, 305
   on teacher attrition, 241
Public service
   compulsory, proposed, 296

   corporate, 108–10
   entitlements, 215, 216
   government, 209
Public works, 272
Pulitzer, Joseph, 7
Pulitzer Prize, 148
Puzo, Mario, 158

Q

Qaddafi, Colonel, 14, 54, 290
QE2 (ship), 11–12
Quality, concept of, 249–64
   American commitment to, 256–58
   convenience and, 252
   costs of, 258–64
   Japanese commitment to, 253–55
   municipal programs, 257

R

Radiation, 289–90
Raiders. *See* Corporate raiders
Railroad barons, 96
Rand Graduate Institute, 244
Rather, Dan, 147, 160, 286
Reader's Digest, 61
Reading
   importance of, 239–41
   U.S. deficiency in, 231–33
Reading is Fundamental, 241
Reagan, Nancy, 145, 166, 172
Reagan, Ronald
   assassination attempt on, 294–95
   character and personality, 164–69, 177–79
   and Chrysler loan, 166
   and education aid, 233
   and Iacocca, 146, 165–66
   Iran/contra issue, 78, 171–73
   management style, 173–74
   political ideology, 290
   popularity, 60, 178, 305, 307
   presidential library, 171
   presidential performance, assessment of, 78, 166–74, 177–79, 287
   presidential style, 268–69, 274, 276–77, 280, 286
   press relations, 152
   public response to, 60, 129, 305
   State of Union message (1985), 292
   Statue of Liberty, 4, 6, 10, 11, 12, 166–67
   trade policies, 156, 176, 192
Reagan administration, 164–79, 288
   business policies, 92
   delegation of authority, 172–74
   drug testing, 207
   education-aid policies, 233
   federal deficit, 206–207, 216–17
   fuel-economy standards, 213
   Iran/contra issue, 171–72
   Regan in, 169–72

Reagan administration *(continued)*
   trade policy, 164–65, 190–92
Reaganomics, 207
Real estate
   farm, 221, 225, 226
   Japanese investments in U.S., 199, 200, 217
Recession. *See* Depression
Regan, Donald, 164, 166–67, 169–72, 173, 174
Reincarnation, 69
Renaissance men, 305–306
Republican Party, 270, 272–73, 279, 280
Retirement
   CEOs' preparation for own, 85–86
   corporate policies, 43–45
   employee pensions and product costs, 261
   Iacocca on, 302–304
Rice, 192–93, 222
Richards, Gil, 39
Robber Barons, 96, 141
Robertson, Pat, 145
Robots, Chrysler's use of, 207–208, 251, 253, 262
Rock-and-roll music, 275
Rogers, Kenny, 14
Rogers, Will, 163
Rohatyn, Felix, 122, 279, 287
Rolls-Royce, 259
Roman Catholic Church, 67–68, 70
Roosevelt, Franklin D., 146, 270–73, 274, 279, 280, 281
Ross, Diana, 60
Rumania, 183
Rumsfeld, Donald, 287
Russia. *See* Soviet Union

S

Safeway Stores, 95
Saint Hugo's Church (Bloomfield Hills, Mich.), 21
Samsung, 236
San Francisco Chronicle, 134
San Quentin prison, 56
Saturn project (GM), 119–20, 177
Scalia, Antonin, 167
Schlesinger, Arthur, Jr., 10, 281
Schlesinger, Arthur, Sr., 281
Schmertz, Herb, 115
Schools. *See* Education; Teachers
Schroeder, Patricia, 291
Scudder, Allen, 123, 129
Seagram's, 105, 283
Seattle Slew (race horse), 60
Securities and Exchange Commission, 272
Service industries
   quality of, 250, 251, 252, 257
   total U.S. workforce, percent of, in, 189–90
Shady, Mrs. (housekeeper), 26
Shays' Rebellion, 225
Shcharansky, Anatoly, 15
Sherman, William Tecumseh, 285

Shultz, George, 164–65
Siegel, Herb, 303
Sign language, 232
Sills, Beverly, 60
Sinatra, Frank, 4, 11, 13
Sloan, Alfred, 121
Small, Isla, 22
Smith, Adam, 177
Smoking
   cigars, 30, 66, 68, 308
   Iacocca's daughters and, 30
   public and, 66, 293
Social conditions (U.S.)
   budget deficit and personal sacrifices, 208–209
   consumption orientation, 213–14
   corporate responsibilities, 108–10
   disadvantaged, aid for, 209, 227–28
   future, 279
   permissiveness, 99–100
   quality of life, 293, 295, 296
   social costs and product competitiveness, 260–64
   standard of living, 218, 256, 260, 291
   world hunger, 227–28
Social Security, 210, 213, 272
Society of Automotive Engineers, 37
South Korea
   press in, 159–60
   trade, 183, 184, 186–87, 188
   workforce, 195–96, 236, 260, 264
Souvenirs, Statue of Liberty, 11
Soviet Union
   Chernobyl accident, 143, 147, 289–90
   space program, 275
   U.S. relations, 287, 290
   U.S. trade, 187
   See also Cold War; Gorbachev, Mikhail
Space program, 250, 275
Spain, 184
Spielberg, Steven, 245
Spitz, Bob, 153
Spock, Dr. Benjamin, 31
Spring Hill (Tenn.) auto plant, 120
Springsteen, Bruce, 11, 14
Star Wars, 244
State Department, 287
State government. *See* Governors
Statue of Liberty–Ellis Island Centennial Commission
   fund raising, 6–9, 59, 62
   gifts to, 59
   Iacocca dismissal from, 3–5, 170
   Iacocca family visits to, 5–6, 39
   Iacocca's feelings for, 5–6
   Iacocca interview on, 153
   Liberty Weekend celebration, 11–14, 39, 155–56, 166
   press and, 153, 155–56
   souvenirs, 11
   as symbol, 5, 6
Steel industry, 96
Stock market
   Chrysler stock performance, 111–12

corporate raiders' impact on, 98, 106
crash (1929), 269, 270
crash (1987), 98, 111–12, 216, 217, 269, 281
public confidence in, 92
Roosevelt's reforms, 272
Suicide, teenagers', 297–99
Suits and claims. *See* Litigation
Supply-side economics, 96, 206
Surgeon General, U.S., 45
Sweden, 184, 299
Switzerland, 212

T

Taiwan, 183, 184, 188, 260
Takeovers. *See* Corporate raiders
Talbott, Michael, 63
Tall Ships, 13
Tanzania, 187
Taxes
  and economic incentives, 292
  export, 211
  and federal deficit, 204–206
  gasoline and oil imports, 216, 217
  Iacocca on, 210–12
  income, 211–12
Taylor, Elizabeth, 14
Teachers and teaching profession
  attrition, 241
  Iacocca in, 304–305
  Iacocca's, 243
  qualifications and training, 242
  salaries, 241–42
  values, imparting, 242–43
Teapot Dome, 159
Teenagers. *See* Young people
Teenage suicide, 297–99
Television
  children and, 236
  Chrysler commercials, 53, 62, 63, 64, 115–17
  Miami Vice, xii, 54, 63–64
  news impact, 147, 160
  parental censorship, 117
  and politics, 145–46
  and U.S. reading levels, 232
Television Information Office, 160
Ten Commandments, 70
Tennessee Valley Authority, 272
Texaco, 133
Thailand, 221–22
Third World. *See* Developing countries
Tiffany, 200
Time magazine, 86, 155, 158, 161
Tire industry, 93–95, 100–101
Tourist industry, 186
Toyota, 197
Trade, 183–201
  agricultural products, 221–22
  deficit, 183–89, 191, 204, 222, 277, 278

deficit solution, 197–98
in Eisenhower era, 275
export taxes, 211
import quotas, 190–91
Japanese dumping, 175–76
Japanese exports and barriers, 184–86, 188–95
"Level Playing Field" speech, 292
Reagan administration and, 164–65, 175–76, 190–92
state policies, 176
surplus, 187
Treasury Department, 287
Trickle-down economics, 97, 206
Trucks. *See* Automotive industry
Truman, Harry, 65, 146, 178, 273–74, 307
Trump, Donald, 78, 245
Tuttle, Holmes, 171
Tutu, Bishop Desmond, 15
TVA, 272
TWA, 96
Twentieth Amendment, 286
Twenty and Eight (cycle theory), 267–82
21 Club, 10

U

UAW. *See* United Auto Workers
Ueberroth, Peter, 287
Underdeveloped countries. *See* Developing countries
Unemployment, 109, 188, 270, 274
Union of Soviet Socialist Republics. *See* Soviet Union
Unisys, 118
United Airlines, 117–18
United Auto Workers, 54, 111
United Jewish Appeal, 283
United Technologies, 85
United Way, 6, 110
Universities. *See* Education; specific institutions
University of Dokkyo, 16
University of Tokyo, 235
Uno Seisakusho Co. Ltd., 186
UNUM, 118
USA Today, 63
U.S. Steel. *See* USX
USX, 91, 96, 118

V

Vaccines, 139
Values
  business, 98–101
  parental guidance in, 28, 30, 242–43
  permissive, 99–100
  and quality, 254
  teachers, 242–43
  young people's, 98–100, 245–46
VCRs, 197, 212
Vermont, 174

INDEX

Vietnam
  refugees, 8
  trade, 187
Vietnam War, 251, 276
Vocational guidance, 236–37, 244–45
Volcker, Paul, 287
Volkswagen, 196
Volunteerism
  corporate, 109–10
  Reagan administration concept, 4, 167

W

Wall Street. *See* Stock market
Wall Street Journal, The, 65, 91, 99, 154, 161
Waldorf-Astoria, 303
Walters, Barbara, 10
Ward, Benjamin, 13
Warranties
  Chrysler, 127, 257
  product quality and, 252
Washington Post, The, 146, 148
Watergate, 141, 159, 251, 276
Welch, Jack, 287
Westheimer, Dr. Ruth, 285
Wheat. *See* Grain
Whirlpool, 257
White, Theodore H. (Teddy), 10
White-collar crime, 99
White Sulphur Springs (W. Va.), 37, 156
Wichita Eagle-Beacon, 148
Wichita State University, 148
Will, George, 164, 169
Wilson, Edith Gault, 172
Wilson, Woodrow, 172, 285–86
Wolper, David, 14, 167
Women, in automotive industry, 84–85

Workers and workforce
  automation and, 251
  educational deficits in, 232–33
  Japanese, 195, 198, 199, 255
  jobs, creating, 272, 291–92
  layoffs, 108–109, 188
  occupational safety and health issues, 261–63
  on-job training, 236
  plant closures, 108–109
  pride in performance, 250–56 *passim*
  remedial education, corporate, 233, 243–44
  service sector, percentage in, 189–90
  South Korean, 195–96, 236, 260, 264
  state competition for jobs, 177
  wages-product price relationship, 260–61
Works Progress Administration, 272
World Bank, 227
World War II, 273, 276, 290, 307
WPA, 272
Wright, Jim, 156
Writing skills
  in business, 87
  young people's, 239–40

Y

Yen, value of. *See* Currency rates
Young people
  aspirations and risk taking, 309–10
  draft/public service proposal for, 295–96
  drugs and alcohol, 295
  and farm crisis, 220–21
  fear of failure, 309
  suicides, 297–99
  values, 98–100
  on Wall Street, 98
  See also Children; Education
Yugo, 183
Yugoslavia, 183